TABLE OF CONTENTS

Radiocarbon 33

Cosmogenic nuclides 61

EDITORIAL

This annual report summarizes the research and development activities of the **Laboratory of Ion Beam Physics** (LIP) and achievements accomplished in 2014. It covers the wide range of fields from fundamental research, over operational issues of the laboratory, to the vast variety of exciting applications of our measurement technologies. We conduct these activities not only in connection to our partners at Paul Scherrer Institut, Empa, Eawag, and other ETH departments, but also with more than 150 external collaborators from Swiss, European, and overseas Universities, from national and international research and governmental organizations as well as for commercial companies. Altogether, the LIP research network is connected to institutions in more than 30 countries. We thank all our external partners for their ongoing support.

LIP's well maintained technical infrastructure is the platform to conduct such a large variety of application projects. Its versatile instrumentation will continue to provide excellent service to internal and external users and contribute significantly to the educational program of ETH.

Among many other major achievements in 2014, the implementation of He stripping into the Aix-Marseille MICADAS system was a special highlight. An overall transmission of almost 50 % was recorded for carbon ions which is equivalent to an ion transport efficiency though the whole instrument of more than 90 %. Equally exciting was the successful integration of permanent magnets (GreenMagnet) into MICADAS systems. This project was conducted in close cooperation with Danfysik A/S and the University of Uppsala is the first institution to benefit from the new technology. It will reduce not only the operational costs of a MICADAS instrument but also simplifies the infrastructure needed to install the system. Green technology has entered the AMS World, which is indicated by the special design of the cover of this report. LIP is going to setup three more MICADAS instruments with GreenMagnets in 2015 before the production will be commercialized and taken over from Ionplus. However, it will be our desire to further develop the ion beam technologies and to make possible a more widespread use of these powerful techniques.

In August 2014 the AMS-13 conference took place in Aix-en-Provence. With 4 out of 23 plenary and a total of 11 out of 121 oral presentations, LIP members have demonstrated their large impact in the AMS community. We are grateful to all LIP staff members who have contributed to continue the LIP activities. Without the excellent scientific, technical and administrative staff, the successful operation of the laboratory would not have been possible.

Hans-Arno Synal and Marcus Christl

THE TANDEM AMS FACILITY

Operation of the 6 MV TANDEM accelerator

Isobar separation for ^{26}Al

Post-accelerator foil stripping for ^{26}Al

OPERATION OF THE 6 MV TANDEM ACCELERATOR

Beam time statistics

Scientific and technical staff, Laboratory of Ion Beam Physics

In 2014, the 6 MV tandem accelerator was in operation for 1250 hours (Fig. 1). Most of the time was devoted to actual measurements or new developments, only less than 5% of the time was used for conditioning and other accelerator related running times. In May 2014 an unscheduled tank opening was necessary because we started to observe terminal voltage (TV) instabilities coming from the generating voltmeter (GVM, Fig. 2). The motor and contacts of the electrodes were repaired.

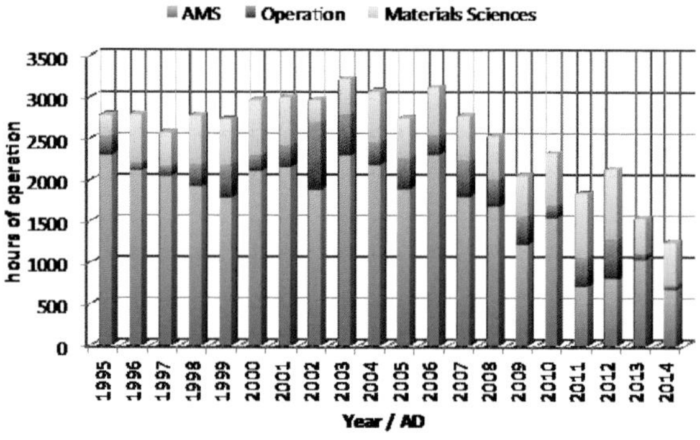

Fig. 1: *Time statistics of the TANDEM operation subdivided into AMS, materials sciences, and service and maintenance activities.*

About 55% of the time was dedicated to AMS measurements. In the past years this time was dominated by ^{10}Be measurements. Due to the transfer of ^{10}Be measurements to the TANDY in 2014 the AMS time went down significantly. The remaining AMS time was split between ^{26}Al and ^{36}Cl routine measurements and tests for isobar separation for $^{26}AlO^-$ and ^{32}Si.

For ^{36}Cl we analyzed 205 samples for geological and hydrological applications as well as for investigations of irradiated materials. The 100 ^{26}Al samples were largely for geological applications and only 10% were coming from irradiated materials. Due to the lower overall efficiency of ^{26}Al the measurement time was longer than for ^{36}Cl.

Fig. 2: *GVM of the Tandem accelerator.*

For materials science applications the beam time increased slightly from 460 to 510 hours. However, the number of analyzed or irradiated samples almost doubled from 1080 to 2090. This is mainly due to numerous measurements of catcher foils which are used to determine angular distributions during thin film deposition. Scans of these foils can be measured semi-automatically and large series of analyses can be obtained very efficiently, i.e. about 12 RBS or ERDA spectra per hour.

The distribution of the terminal voltage (Fig. 3) clearly shows the two main application fields, material sciences operating at lower TV and AMS at high TV. In the future that pattern will likely change because in 2014 a new beam line was built at the 0° port of the HE AMS magnet for MeV SIMS which will also run at higher TV.

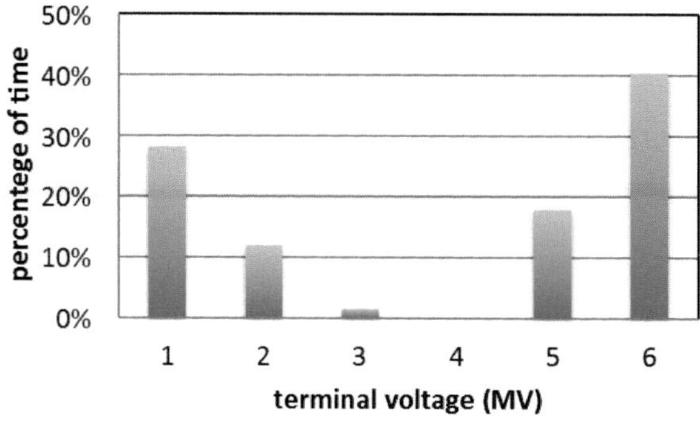

Fig. 3: Time distribution of terminal voltages.

ISOBAR SEPARATION FOR ^{26}Al

Comparison of foil degrader and gas-filled magnet

C. Vockenhuber, K.-U. Miltenberger, H.-A. Synal, M. Suter, R. Gruber, T. Mettler

We investigated the suppression of the isobar pair ^{26}Al and ^{26}Mg at energies available at the 6 MV EN tandem accelerator for improved efficiency of AMS measurements using AlO^-. When injecting this molecule, the ion energy of ^{26}Al at 6 MV and charge state 6+ is 38 MeV.

In a typical sample we can expect a Mg content at the level of ppm; thus for measurements of ^{26}Al/^{27}Al ratios down to 10^{-15}, ^{26}Mg must be suppressed by additional 8 orders of magnitude (accounting for a suppression of 10 due to the isotopic abundance of ^{26}Mg). The gas ionization detector provides the main suppression at a factor of 10^5 (Fig. 1). The remaining 10^3 can be achieved by either the degrader foil technique or the gas-filled magnet (GFM) method.

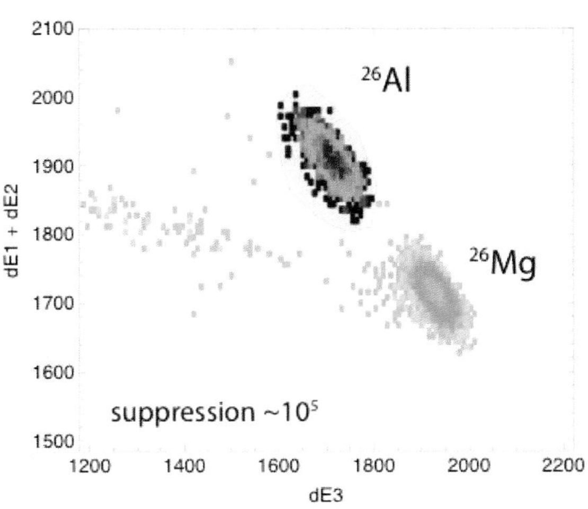

Fig. 1: *Isobar separation in the gas detector.*

Both methods were investigated at the 180° GFM at the end of the AMS beam line. For the degrader method we installed a 1500 nm SiN foil at the entrance of the GFM running without gas, based on energy-loss and straggling measurements performed in an earlier experiment [1]. The scan in Fig. 2a shows the separation of ^{26}Al and ^{26}Mg. The resulting suppression factor of 500 is slightly higher than calculated from the earlier measurements [1].

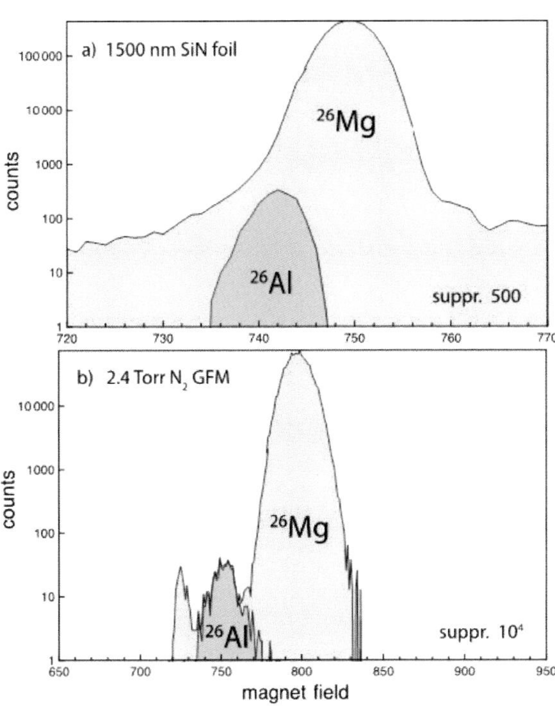

Fig. 2: *Magnet scans with the 1500 nm SiN degrader foil (a) and 2.4 Torr of N_2 in the GFM (b). ^{26}Al counts appear at lower magnetic fields than ^{26}Mg, allowing for a suppression of the ^{26}Mg count rate.*

For the GFM method we installed a thin 100 nm SiN entrance window and filled the magnet chamber with a few Torr of N_2. The separation of ^{26}Al and ^{26}Mg is then based on the mean charge (Fig. 2b). In the direct comparison of both methods the GFM results in more symmetric peak shapes and thus a higher suppression factor (10^4). In order to fully utilize the new measurement procedure we need a larger detector that can accept the wide beams after the broadening due to the degrader foil or the gas in the GFM.

[1] Miltenberger et al., LIP 2014 annual report

POST-ACCELERATOR FOIL STRIPPING FOR ^{26}Al

Evaluation of ^{26}Mg suppression for ^{26}Al measurements using AlO^-

K.-U. Miltenberger, C. Vockenhuber, T. Mettler

Most AMS measurements of ^{26}Al rely on the use of the negative atomic ion Al^- in order to completely suppress the interfering isobar ^{26}Mg. However, the Al^- currents are at least one order of magnitude lower than those obtained when extracting the molecular ion AlO^-, and thus the overall precision and detection limit are restricted by counting statistics. The difficulties of using AlO^- arise from the much more abundant ^{26}Mg introduced via MgO^-. Separation of ^{26}Mg from ^{26}Al can be achieved by making use of the different energy loss in post accelerator stripper foils and adding a succeeding energy filter (e.g. magnet). To evaluate this method, energy loss and straggling measurements at 32 and 38 MeV were performed for both ions at the 90° HE magnet of the 6 MV Tandem with silicon nitride (SiN) foils of 500 to 2000 nm thickness.

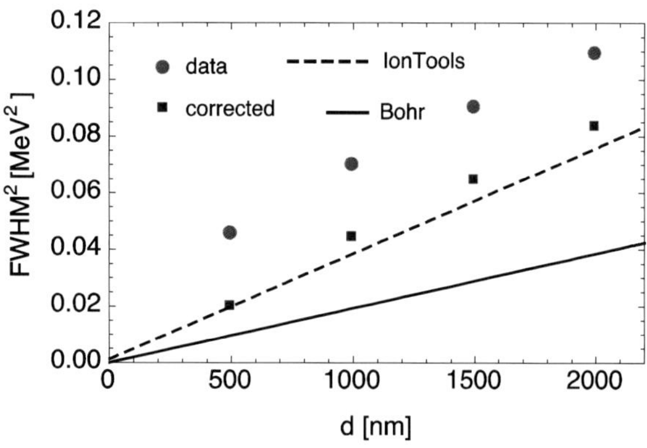

Fig. 1: *FWHM^2 vs. thickness of SiN foil ($^{26}Mg^{6+}$).*

While energy loss predictions by SRIM and IonTools [1] are reasonable, the measured values of the relevant energy loss difference between ^{26}Al and ^{26}Mg were consistently lower than both estimates. In the case of energy straggling (FWHM), the Bohr straggling formula underestimates the straggling significantly, while IonTools estimates (based on the Yang formula) match the measurements (corrected for beam energy spread and measurement resolution) quite well (Fig. 1).

The achievable suppression levels of ^{26}Mg depend on energy, SiN foil thickness and cuts in the energy distribution which also influence the acceptance levels of ^{26}Al. Additionally, the beam losses due to angular straggling and the stripping yield were taken into account. A modified peak model accounting for the asymmetric peak and low energy tail of the energy distribution was employed to calculate realistic suppression and acceptance factors.

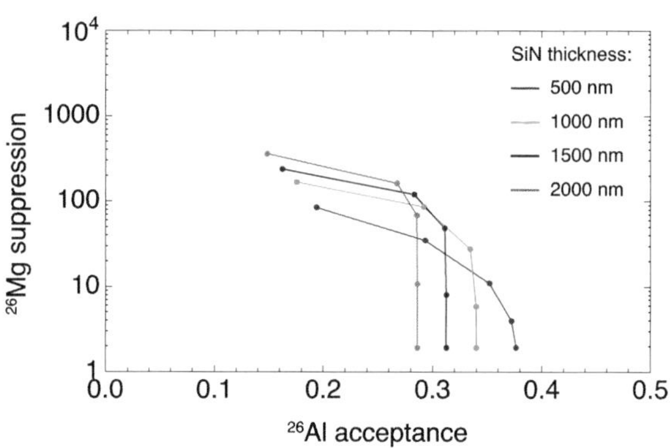

Fig. 2: *$^{26}Mg^{6+}$ suppression vs. $^{26}Al^{6+}$ acceptance for different SiN foils and energy cuts at 38 MeV.*

Depending on the SiN foil thickness and the cut in the energy distribution the suppression and acceptance can be optimized (Fig. 2). With a 1500 nm SiN foil the ^{26}Mg suppression factor is about 200 at an ^{26}Al acceptance of around 30%. The main loss of ^{26}Al acceptance comes from the charge state yield (36%); losses from angular scattering depend on the size of the detector. The achieved ^{26}Mg suppression together with the separation in the gas ionization detector should be sufficient to exploit the higher ^{26}Al current from the extraction of AlO^-.

[1] www.srim.org; R.A. Weller, AIP CP 475 (1999) 596

THE TANDY AMS FACILITY

Activities on the 0.6 MV TANDY in 2014

Towards a compact multi-isotope AMS system

^{26}Al measurements at the Tandy facility

Intercomparison of ^{236}U in seawater samples

ACTIVITIES ON THE 0.6 MV TANDY IN 2014

Beam time and sample statistics

Scientific and technical staff, Laboratory of Ion Beam Physics

In 2014, the multi-isotope facility TANDY accumulated almost 2500 operation hours for routine AMS measurements and for technical developments. More than 2200 AMS samples were analyzed for various nuclides and different applications.

Fig. 1: *Temporary replacement of the Pelletron system with a MICADAS-type accelerator.*

The technical highlight of 2014 was the replacement of the NEC Pelletron system by a MICADAS type accelerator housing running at 250 kV (Fig. 1). For two months this prototype setup remained installed, its performance for AMS analyses of ^{26}Al, ^{41}Ca, ^{129}I, and the actinides was tested – with encouraging results.

In summary, about 10% of the total annual operation time was spent for technical developments (Fig. 2), also including the development of new, high performance gas ionization detectors. The majority (90%) of operating time was again allocated for routine AMS analyses. During half of this time ^{10}Be was measured, the remaining half equally splits between ^{129}I and the actinides (Fig. 2).

About ½ of all AMS samples (1100) were analyzed for ^{10}Be (Fig. 3), most of them for ice core studies and in-situ dating. More than ¼ of all samples (600) were measured for ^{129}I involving environmental monitoring (soils, rain water, air filters), and oceanographic studies.

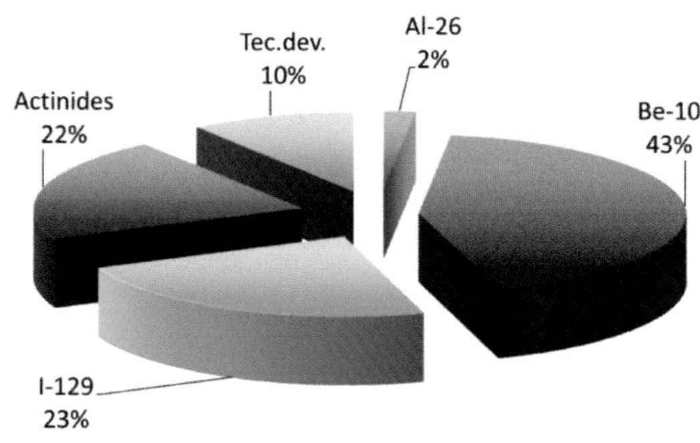

Fig. 2: *Relative distribution of the TANDY operating time for the different radionuclides and activities in 2014.*

More than 500 actinide samples (22%, Fig. 3) were analyzed for applications in oceanography (^{236}U and Pu), for environmental monitoring programs (e.g. Fukushima), and to develop new techniques for human bioassay studies using U, Pu, Am, Cm, and Cf isotopes. Finally, few samples were measured for ^{26}Al, in order to establish a new absorber technique for AMS of ^{26}Al in the 2+ charge state.

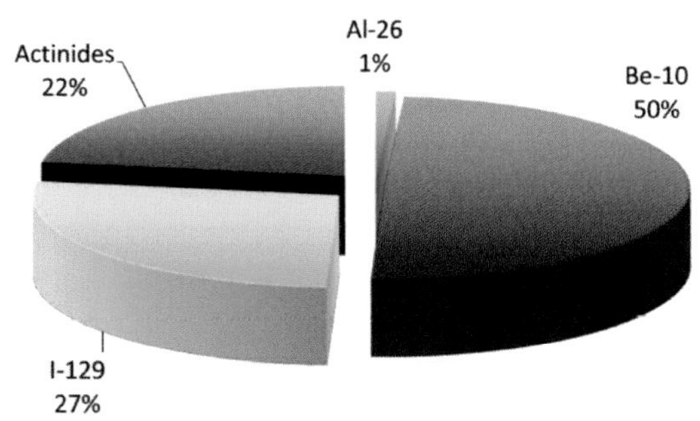

Fig.3: *Relative distribution of the number of AMS samples measured for the various radionuclides in 2014.*

TOWARDS A COMPACT MULTI-ISOTOPE AMS SYSTEM

Performance of the proof of principle experiment

S. Maxeiner, A. Herrmann, A. Müller, M. Suter, M. Christl, C. Vockenhuber, H.-A. Synal

The MICADAS radiocarbon AMS system is well-known for its compact footprint and vacuum insulated accelerator. At LIP, developments are going on to develop a compact multi-isotope AMS system based on a MICADAS-type accelerator. In a first proof of principle experiment, such a prototype system was assembled by replacing the existing TANDY Pelletron accelerator with a modified MICADAS acceleration unit running at 250 kV (Fig. 1).

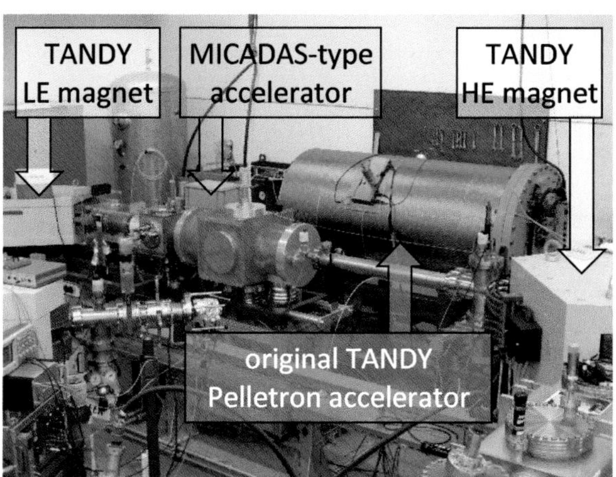

Fig. 1: *The TANDY Pelletron accelerator (blue, >2 m in length) was exchanged with a modified MICADAS type accelerator (yellow, ca. 1 m).*

To ensure correct ion optical focusing of the new setup the ground and terminal electrodes of the accelerator had to be redesigned. Furthermore, a new stripper tube for the use with He was designed in order to optimize vacuum conditions, beam transmission and background suppression. Helium gas provides high stripping yields and less angular straggling than stripping with heavier gases [1]. The much smaller tube diameter of 3.2 mm compared with 8 mm of the original TANDY stripper reduced the gas outflow by a factor of two to three. Nevertheless the optical beam transmission could be improved or at least kept the same by increasing the opening angle of the new stepped stripper tube from 16 mrad (TANDY) to 24 mrad.

First AMS measurements of ^{26}Al, ^{41}Ca, ^{129}I and ^{236}U with the prototype system revealed a good performance for all nuclides when compared with the respective routine AMS setup at ETH Zurich (Tab. 1). Considerable improvement was found for actinide measurements. Here, the abundance sensitivity could be enhanced by a factor of 10 compared to the routine setup. This can be explained by the better vacuum and the new stripper, which acts as an additional aperture.

Nuclide	^{26}Al		^{41}Ca		^{129}I		^{236}U	
Acceleration voltage / kV	4500	250	500	250	300	250	300	240
Stripper gas	Ar	He	He	He	He	He	He	He
Post stripping charge state	3+	2+	2+	2+	2+	2+	3+	3+
Stripper trans-mission / %	20	51	50	49	43	51	35	35

Tab. 1: *Performance parameters for the measured nuclides (first column: routine AMS setups [2], second column: this experiment).*

Further improvements of the gas feeding system and the HV feed-through should in principle allow to apply voltages up to 300 kV. This would improve the beam transport on the high energy side and the subsequent particle detection.

[1] C. Vockenhuber et al., Nucl. Instr. & Meth. B 294 (2013) 382

[2] M. Christl et al., Nucl. Instr. & Meth. B 294 (2013) 29

^{26}Al MEASUREMENTS AT THE TANDY FACILITY

Performance below 500 kV for the 2+ charge state

A.M. Müller, M. Christl, J. Lachner, S. Maxeiner, H.-A. Synal, C. Vockenhuber, C. Zanella

For some time helium is used as stripper gas at the ETH 500 kV Tandy system, which enhanced the accelerator transmission for all the radionuclides routinely measured at this facility [1]. Investigations with aluminum showed that a transmission of more than 42 % is achievable at the Tandy for the 2+ charge state. But the *m/q* ambiguity $^{13}C^{1+}$ can pass the spectrometer at intensities up to the pA range, which can't be handled with gas or solid state detectors.

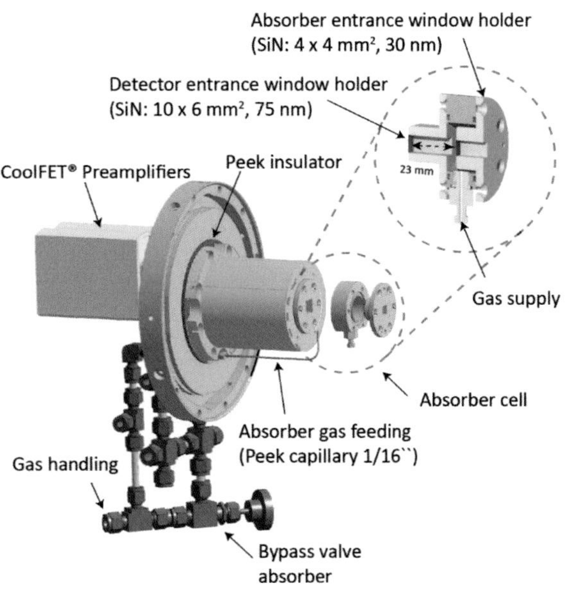

Fig. 1: *Perspective view of the new GIC with the dedicated absorber cell [3].*

In order to stop the ^{13}C interference before entering the detector first tests with a simple absorber cell in front of the ETH ΔE-E$_{res}$ gas ionization chamber (GIC) were successfully performed [2]. For this reason a more sophisticated GIC design was built with an dedicated absorber cell, which can be filled independently from the detector volume (Fig. 1). With this design an ^{26}Al overall transmission (LE side into the GIC) of 40% and 29% could be achieved at 500 kV and 300 kV respectively, while the ^{13}C interference was completely suppressed by the gas absorber. ^{26}Al/^{27}Al ratios of $<10^{-14}$ were obtained from aluminum dummy samples. A comparison measurement performed at the TANDY and the ETH 6 MV Tandem facilities of 17 real samples covering a ^{26}Al/^{27}Al range between 10^{-14} - 10^{-10} showed an good correlation between the two systems, although some low level samples were measured higher at the Tandy system (Fig. 2).

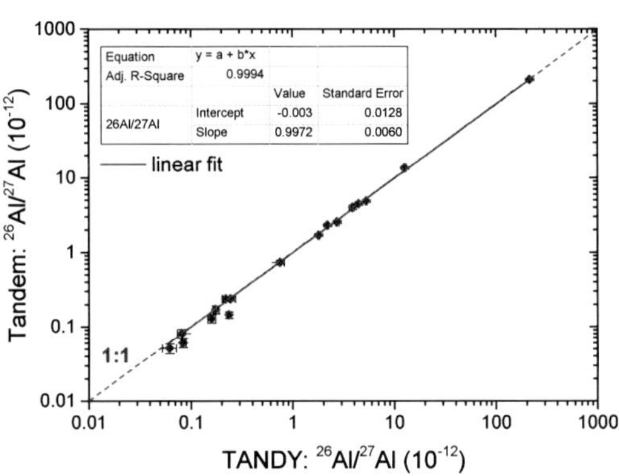

Fig. 2: *Comparison of 17 samples measured at the Tandy system and the ETH 6 MV Tandem facility [3].*

This measurement demonstrates the great potential of the absorber technique for ^{26}Al in the 2+ charge state below 500 kV. Further comparisons and investigations on Al$^-$ the ion source output, which was limited to a few hundreds of nA, have to be performed.

[1] C. Vockenhuber et al., Nucl. Instr. & Meth. B 294 (2013) 382

[2] J. Lachner et al., Nucl. Instr. & Meth. B 331 (2013) 209

[3] A.M. Müller et al., Nucl. Instr. & Meth. submitted

INTERCOMPARISON OF ^{236}U IN SEAWATER SAMPLES

First intercomparison exercise between CNA-Sevilla and ETH-Zurich

E. Chamizo[1], M. López-Lora[1], N. Casacuberta, M. Christl, M. Villa[2]

The determination of ^{236}U ($T_{1/2} = 23 \cdot 10^6$ y) in oceanographic samples has become an important application field in AMS. The low reported $^{236}U/^{238}U$ atom ratios for seawater samples (ranging from 10^{-12} to 10^{-6} [1]) can only be assessed by this technique. In the last years, the upgraded version of the 600 kV TANDY AMS facility at the ETH with an additional high energy (HE) magnet has been established in this field. Recently, the 1 MV compact AMS facility at the *Centro Nacional de Aceleradores* (CNA, Seville, Spain) has proved its potential for the determination of anthropogenic ^{236}U [2]. In contrast to the TANDY, the CNA facility has only one HE magnet and thus a lower abundance sensitivity.

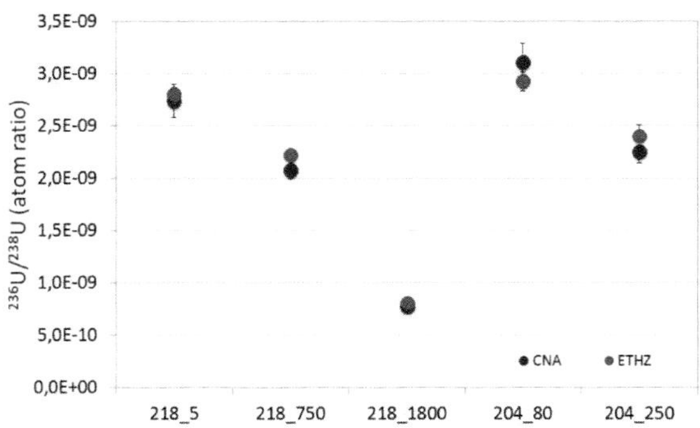

Fig. 1: *ETH and CNA results of $^{236}U/^{238}U$ atom ratios for the 5 studied seawater samples. The first number designates the sampling station, the second one the water depth in meters.*

In this study, the applicability of the CNA AMS facility to ^{236}U studies in seawater samples has been explored. For this purpose, five seawater samples from the Arctic Ocean provided by ETH have been analyzed at CNA. Different water depths from two sampling stations were chosen in order to have a broader range for the $^{236}U/^{238}U$ atom ratio (Fig. 1). Aliquots of 5 l were used except for the deepest sample at station 218, where 10 l were processed. UTEVA® cartridges were used for the final separation of the uranium fraction from the first iron hydroxide precipitate [2].

CNA results for the $^{236}U/^{238}U$ atom ratios are compared with the values obtained at ETH for the same samples (Fig. 1). There is an excellent agreement between both sets of results, also for the sample with the lowest value (of about $8 \cdot 10^{-11}$, sample 218_1800). This ratio is only a factor of 20 higher than the abundance sensitivity achieved at the CNA AMS facility for $^{236}U/^{238}U$. Additionally, a very good agreement has been obtained for the ^{236}U atom concentrations for the four studied samples that were spiked with ^{233}U, compiled in Tab. 1 [1, 2].

This is the first intercomparison exercise for ^{236}U performed in seawater samples.

	^{236}U conc. (10^6 atoms/kg)	
Sample	CNA	ETH
218_5	21 ± 1.3	21.9 ± 0.9
218_750	18.6 ± 0.9	20.1 ± 0.8
204_80	24.8 ± 1.7	23.9 ± 0.9
204_250	19.7 ± 1.1	19.6 ± 1.1

Tab. 1: *ETH and CNA results of ^{236}U atom concentrations in the studied seawater samples that were spiked with ^{233}U.*

[1] N. Casacuberta et al., Geochim. Cosmoschim. Acta 133 (2014) 34
[2] E. Chamizo et al., Nucl. Instr & Meth. B (2015) submitted

[1] *Centro Nacional de Aceleradores, Universidad de Sevilla, Spain*
[2] *Departamento de Física Aplicada II, Universidad de Sevilla, Spain*

THE MICADAS AMS FACILITIES

RADIOCARBON MEASUREMENTS ON MICADAS IN 2014

Performance and sample statistics

Scientific and technical staff, Laboratory of Ion Beam Physics

In 2013, our MICADAS AMS facility was operated on 330 days with 5500 hours spent on measuring samples, thus reaching its full capacity. In order to further increase the measurement capacity in 2014, the measurement protocol at the MICADAS was optimized for higher ion currents, thus reducing the required measurement time for a single sample. Routine measurements of solid graphite samples are now performed at $^{12}C^{-}$ currents of about 80 μA from the original 40-60 μA. Two instead of one magazine of 22 samples can now be measured in a day.

On 310 days in 4500 hours more than 8000 samples were measured in 2014, an increase of more than 20% over 2013! While the increase on solid graphite was about 15%, we analyzed about 30% more gas samples (Fig. 1). Usage of carbonate handling system (CHS) installed in 2013 to analyze the headspace of septa-sealed vials contributed more than 600 carbonate samples [1] and 500 contamination swipes [2]. Carbonate analysis capabilities were extended by the addition of a new direct injection interface [3].

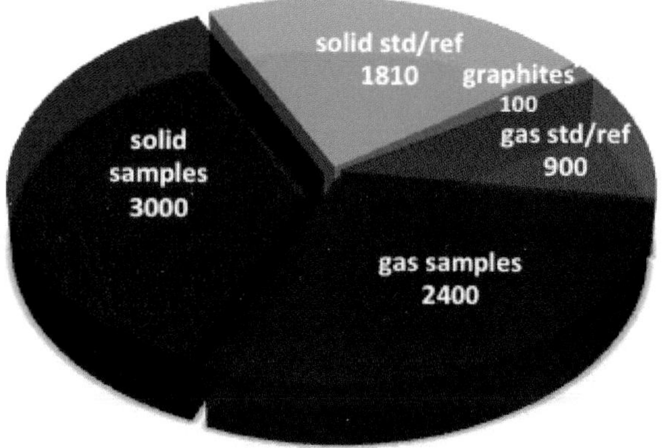

Fig. 1: *Samples measured on MICADAS in 2014.*

Graphite samples and standards prepared at ETH Zurich are in blue, samples graphitized outside ETH are in gray. Red indicates samples measured with the gas ion source.

2500 samples were measured for our partner institutions which were closely followed by about 2000 commercial samples. About 800 samples were measured for internal projects.

While we ran significantly more samples than in the previous year, we still found some time to make progress in automatizing and extending the gas interface system. Fully automated online radiocarbon measurements of samples combusted in the elemental analyzer are now possible [4] and we expect first routine measurements in early 2015. Also, the stable isotope mass spectrometer was integrated with the gas handling system and allowed for the first time a precise simultaneous measurement of $\delta^{13}C$ and radiocarbon values [5]. The goal for 2015 will be to routinely run precise $\delta^{13}C$ measurements in parallel with radiocarbon measurements on gas samples. Also, a second MICADAS system shall be installed to extend measurements capacities at ETH Zurich and LIP in particular. This will prepare us for an expected increase in radiocarbon samples, mostly from our partner institutions at ETH.

[1] L. Wacker et al., LIP annual report (2014)23
[2] C. McIntyre et al., LIP annual report (2014) 50
[3] M. Seiler et al., LIP annual report (2014) 25
[4] S. Fahrni et al., LIP annual report (2014) 26
[5] C. McIntyre et al., LIP annual report (2014) 24

THE GreenMICADAS PROJECT

The first AMS Instrument with permanent magnets

H.-A. Synal, S. Maxeiner, M. Seiler, F. Bødker[1], L.O. Baandrup[1], A. Baurichter[2], M. Salehpour[2], G. Possnert[2]

In general, dedicated radiocarbon dating AMS instruments are best suited to implement permanent magnet technologies. Designed for the analyses of carbon isotopes only, the required magnetic fields of the two spectrometer magnets are fixed and need only be varied over a small field range during system the tuning process. So far, permanent magnets have not been successfully implemented in AMS because the temperature dependence of the permanent magnetic materials have prevented stable enough operating conditions. To overcome this problem, Danfysik A/S has developed a new design of the MICADAS-type LE- and HE-magnets using flux shunting [1,2,3].

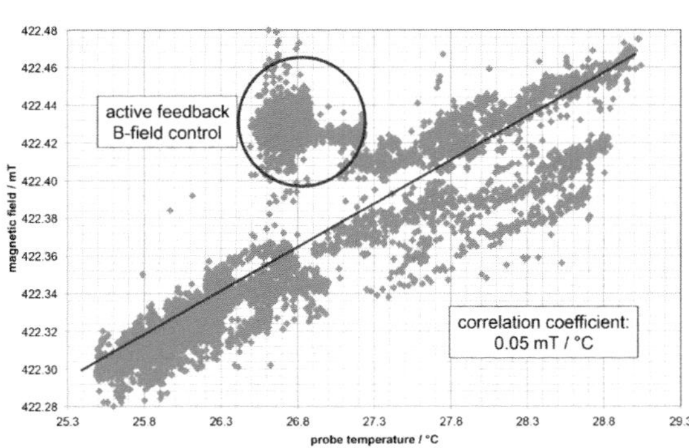

Fig. 1: *Long-term observation of the field stability of the GreenMagnet using Group3 hall probe for field monitoring.*

The moderate central field of the LE magnet (0.4 T) allows a design where the permanent magnets (high grade NdFeB with a typical remanence of 1.4 T) are placed between the iron poles. During factory tests of the LE-magnet a thermal stability of better than 20 ppm/°C was measured using NMR field probing. In Zurich, we made a long-term field monitoring using a Group3 hall probe. Data taken during a period of more than half a year showed a field temperature relationship of about 50 ppm/°C (Fig. 1). But, hall probe measurements are less accurate because the measurement procedure itself has temperature dependence of the order of the magnetic field variations. However, we have implemented a feedback loop to stabilize the field to the probe reading (red circle in Fig. 1). In any case, our results show the temperature compensation is by far sufficient to maintain a stable operation of an AMS system.

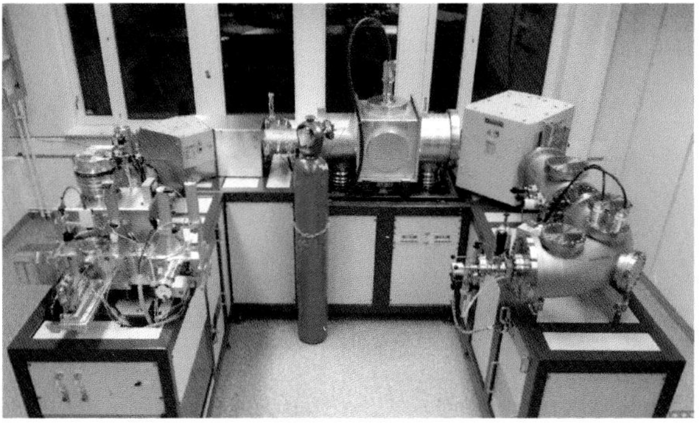

Fig. 2: *The GreenMICADAS AMS spectrometer installed at the Angström laboratory in Uppsala.*

The HE-magnet caused more difficulties and a careful design of the fringing field was needed to obtain the appropriate ion optical focusing performance. The system has been completed in September 2014 and is now installed and fully operational at the Ångström Laboratory at Uppsala University. Operation of the entire AMS system requires only 8 kW electrical power; additionally there is no need for a cooling system to dissipate the heat produced inside the coils of conventional magnets.

[1] F. Bødker et al., PAC (2013) 3511
[2] S.H. Kim and C. Doose, PAC (1997) 3229
[3] H.D. Glass et al., PAC (1997) 3262

---

[1] *Danfysik A/S, Taastrup, Denmark*
[2] *Ångström Laboratory, Uppsala University, Sweden*

THE AixMICADAS PROJECT

Improving system performance by using He stripping

H.-A. Synal, S. Maxeiner, M. Seiler, L. Wacker, S. Fahrni , E. Bard[1,2], Y.Fagault[1], T. Tuna[1], L. Bonvalot[1]

The latest version of a MICADAS-type, dedicated ^{14}C AMS spectrometer has been installed at the CEREGE campus of the Aix-Marseille University (Fig. 1).

Fig. 1: *The AixMICADAS spectrometer installed at CEREGE in Aix-en-Provence, France.*

The system incorporates several new features compared to previous MICADAS instruments. In particular, the gas stripper has been converted to enable operation with He stripper gas. The design of the MICADAS acceleration unit requires a supply of the stripper gas from ground potential. Pressurized He is not as highly electrically insulating as N_2 or Ar, which are commonly used as stripping media in low energy AMS. To prevent electrical discharges, the pressure of the He gas inside the capillary feeding the stripper tube was increased to more than 2.5 MPa. By using He, the conductance of the apertures confining the stripper housing is higher causing an increased outflow of stripper gas into the acceleration section. However, the angular straggling of the ion beam in He is significantly reduced. Careful adaptation of the geometry of the aperture at the exit of the stripper housing to the true phase space volume of the ion beam has improved vacuum conditions.

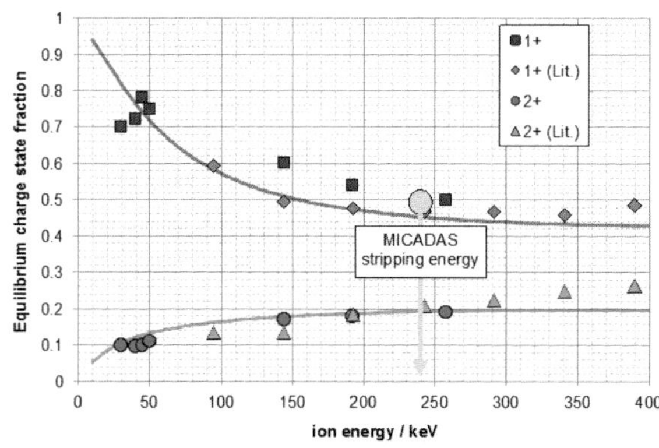

Fig. 2: *Charge state yields of carbon 1+ and 2+ in He under equilibrium conditions.*

At the stripping energy of the MICADAS (Fig. 2) the yield of the 1+ charge state in He is close to 50%. Based on the observed overall transmission of approx. 47% the ion optical transmission is at least 90%, which can be regarded as a precondition for stable and reproducible measurements. Under routine operation an overall uncertainty of 2‰ is reached on modern samples of about 1 mg of carbon.

Direct measurements of geological graphite indicate a machine background equivalent to an age of 68'000 yr BP [1]. Hence, dating performance at the AixMICADAS is limited solely by the ^{14}C contamination of samples in the field and during preparation of the target material. The system is further equipped with a gas interface, accepting CO_2 from the elemental analyzer and the automated carbonate hydrolysis unit. The direct coupling is fully automated and allows unattended operation.

[1] E. Bard et al., AMS-13 proceedings

---

[1] *Aix-Marseille University, Aix-en-Provence, France*
[2] *Collège de France, Paris, France*

REARRANGED MyCADAS

New setup for improved differential pumping and gas detector

M. Seiler, S. Maxeiner, L. Wacker, H.-A. Synal

At the 45 keV AMS system MyCADAS the analyzing magnet and the electrostatic analyzer (ESA) were swapped for mounting a gas ionization chamber (GIC), resulting in a setup similar to the proof-of-principle experiment [1]. Due to the focal length of the analyzing magnet, it can be placed about 30 cm further away from the stripper tube than the ESA in the previous setup resulting in a slightly larger system footprint (Fig. 1).

***Fig. 1:** Picture of the modified MyCADAS with a gas ionization chamber for single ion detection.*

The increased distance between analyzing magnet and the stripper allows an improved implementation of the differential pumping to achieve a vacuum level of 2x10^{-6} mbar even with an increased stripper tube diameter of 3.2 mm instead of 3 mm. The exit of the stripper section was redesigned and completely separates the spectrometer from the stripper section. The only connection is given by an opening for the ion beam so that the helium outflow is minimized (Fig. 2). An additional turbo pump is placed directly after this tube.

The exchange of the ESA and the analyzing magnet leads to a positioning of the Faraday cups in the vacuum chamber between these two elements. Since the Faraday cups are not placed in the same vacuum chamber as the ^{14}C detector anymore, the system is a bit less stable against drifts of filter elements. Ion optically the spectrometer makes a 1:1 mapping from the stripper tube to the beam waist after the ESA. In this position the GIC is installed for ^{14}C detection [2]. Compared to the electron multipliers used before, the GIC has similar detection efficiency, no dark counts and is more resistant to high count rates.

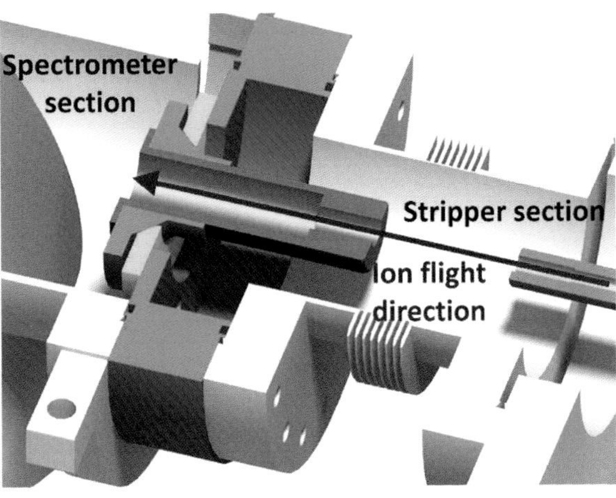

***Fig. 2:** Differential pumping system at the exit of the stripper section. The connection of the inner stripper volume to the spectrometer is only given by a small connection tube (purple).*

With the new arrangement the stripper transmission for ^{12}C ions is almost 40%. The detection efficiency of about 60% for ^{14}C ions is given by the efficiency of the detector and beam losses at the apertures in the ESA. The background level corresponds to a radiocarbon age of 40'000 years.

[1] H.-A. Synal et al., Nucl. Instr. & Meth. B 294 (2013) 349

[2] A. Müller et al., LIP Annual Report (2014) 30

FIRST ^{14}C-SCANS ON STALAGMITES BY LASER ABLATION

Laser Ablation-AMS reveals prominent ^{14}C features in carbonate records

C. Welte[1], L. Wacker, B. Hattendorf[1], J. Koch[1], M.Christl, H.-A. Synal, D. Günther[1]

When applying laser ablation (LA) to carbonates a high proportion of the ablated $CaCO_3$ is converted into CO_2 [1]. The ^{14}C content of this CO_2 can be analyzed with gas ion source (GIS)-AMS [2]. For rapid online ^{14}C analysis of carbonates a LA-setup [3] was developed that allows to perform different scanning modes with an ArF excimer laser (193 nm) on large natural samples (maximum dimensions: 15 x 1.5 x 2.5 cm^3) like stalagmites or corals. More than 10 cm can be scanned within less than one hour with this new technique.

Fig. 1 depicts the radiocarbon bomb spike in a stalagmite once derived from conventional micromilling (black squares) compared with data from LA analysis (red circles). Continuous scans (scanning velocity of 140 µm/s, laser repetition rate of 200 Hz and a fluence of 1-1.5 J/cm^2) from top to bottom and in the opposite direction were performed. The mean of these two scans was taken to increase the precision. The data were divided into subsamples equal to 30 s of data acquisition resulting in a spatial resolution of ~ 100 µm and a precision of 3-4 %. The LA data could confirm the abrupt onset of the bomb peak.

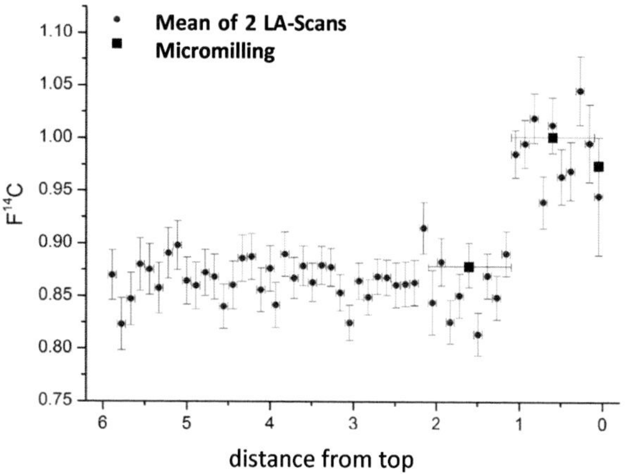

Fig. 1: *Bomb spike in a stalagmite analysed by conventional micromilling (black) and LA (red).*

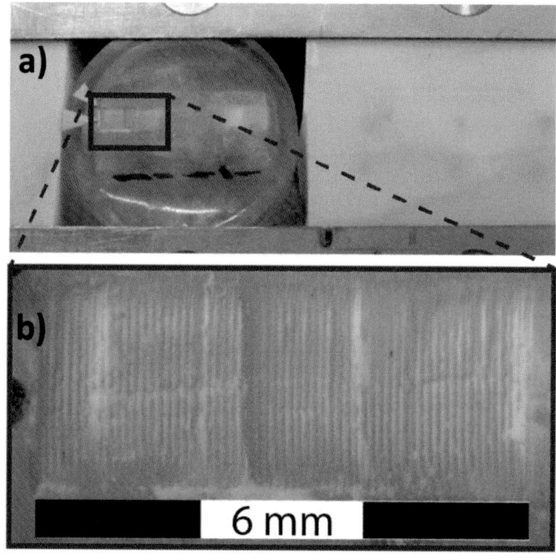

Fig. 2: *Stalagmite sample in the sample holder (a) and enlarged section of the laser track (b).*

Fig. 2a depicts the measured piece of stalagmite embedded in epoxy resin and placed in the sample holder (metal frame) together with a marble blank sample. An enlarged section of the laser tracks is shown in Fig. 2b.

LA has proven to be a powerful sampling technique for ^{14}C-analysis that combines short analysis times with high spatial resolution in the order of 100 µm or smaller. This allows the quick detection of prominent features in carbonate records like the bomb spike or growth stops. Future work will cover modifications of the setup to make the CO_2 transport into the GIS more efficient allowing to reach a higher measurement precision.

[1] L. Wacker et al., NIM B 294 (2013) 287
[2] M. Ruff et al., Radiocarbon 49 (2007) 307
[3] C. Münsterer et al., Chimia 68 (2014) 215

[1] *D-CHAB, ETH Zurich*

ADVANCED GAS MEASUREMENTS OF FORAMINIFERA

Removal and analysis of carbonate surface contamination

L. Wacker, S. Fahrni, M. Moros[1]

Most laboratories require more than 3 mg of carbonate material for a single radiocarbon analysis by AMS. However, with the gas ion source of MICADAS we need less than 1 mg of carbonate: CO_2 is liberated from carbonates in septum sealed vials by acid decomposition, subsequently collected on a zeolite trap and transferred to a syringe from where it is fed into the ion source [1]. This fast handling of small samples provided us with the opportunity to measure very small leach fractions of carbonate samples. While most carbonate samples at ETH have been measured without this treatment, it is sometimes necessary to remove surface contaminants with acid leaches. This is particularly important for old foraminifera [2].

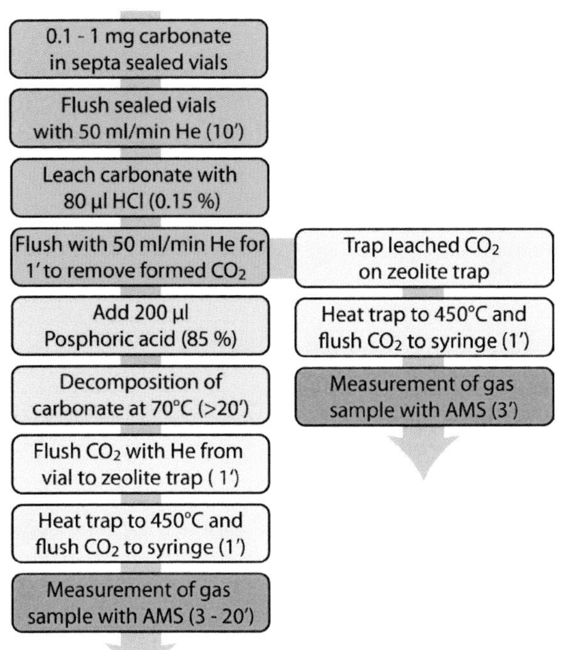

Fig. 1: *Procedure for analyses of foraminifera samples including sample leaching.*

An acid leach step was therefore added for gaseous measurements of small foraminifera samples (Fig. 1). Only 80 µg of carbonate (10 µg C) is removed with 80 µl of 0.15% HCl in septa sealed vials and the formed CO_2 is measured. The remaining carbonate is then completely dissolved and measured in a second step.

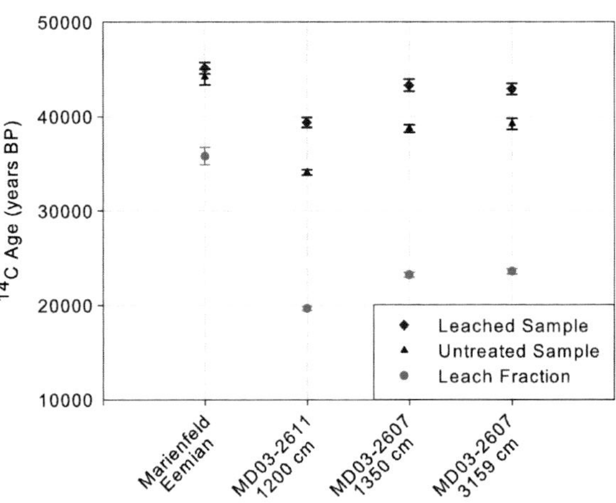

Fig. 2: *^{14}C concentrations of the leach fraction (blue) and the leached foraminifera sample (red) are compared with the untreated sample (black) and a carbonate blank (IAEA-C1).*

Fig. 2 illustrates the difference of the leach and the remaining fraction for old samples compared to the analysis of untreated samples. While the Eemian foraminifera sample from the Baltic Sea is hardly influenced by surface contamination, samples from the Australian coast show a non-negligible surface contamination. The measurement of the leach fraction in addition to the leached sample ensures a good quality control and allows us to identify problematic samples.

[1] L. Wacker et al., Nucl. Instr. & Meth. B 294 (2013) 307

[2] M. Schleicher et al., Radiocarbon 40 (1998) 85

---

[1] *IOW Rostock, Germany*

PRECISE ^{13}C AND ^{14}C GAS MEASUREMENTS

Integration of a stable isotope mass spectrometer

C. McIntyre, S. Fahrni, L. Wacker, T. Eglinton[1]

Samples analyzed for radiocarbon in global carbon cycle studies require ^{13}C measurements with a precision of better than 0.1‰ for interpretation of individual processes. ^{13}C measurements on the MICADAS system at ETH Zürich are used for real-time correction of ^{14}C ratios but their overall accuracy and precision is insufficient for our biogeochemical applications. A new Isoprime VisION stable isotope mass spectrometer (IRMS) has been integrated with an elemental analyzer (EA), the gas interface system (GIS) and the gas ion source of the MICADAS AMS system (Fig. 1). The overall aim is to provide automated and high throughput, high precision ^{13}C and ^{14}C analysis of bulk materials through to individual compounds.

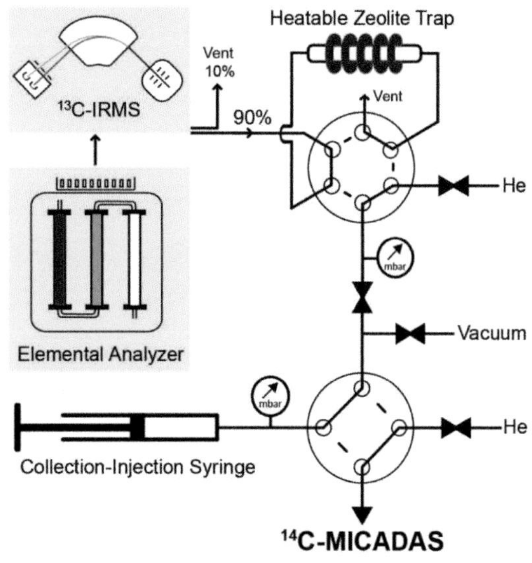

Fig. 1: *EA-IRMS-AMS system schematic.*

Samples containing 50-100 µg of carbon are combusted by the EA to produce CO_2. The EA flow is split internally by the IRMS with ~ 90% of the gas flow transferred to the gas interface system. CO_2 is trapped on a zeolite trap before transfer with helium to the AMS ion source via a syringe pump. Total analysis time is 15 minutes per sample.

To date, work has included designing a compact format, optimizing the flow path and programming software [1]. The system can run in an automatic supervised mode and has been tested with standard reference materials. ^{14}C Fm values fell within error of their consensus values and a precision of better than 1% for a modern sample was readily achievable. Background was measured at 44'000 years for a radiocarbon dead compound. ^{13}C values with a precision of better than 0.1 permil for n = 6 standards were also readily achievable (Fig. 2).

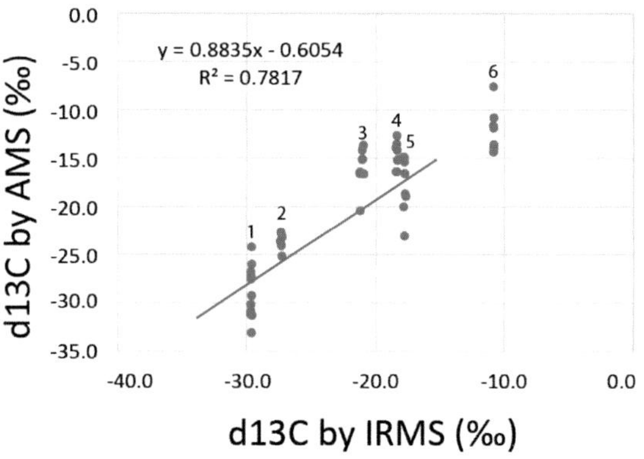

Fig. 2: *Comparison of δ^{13}C measurements by AMS and IRMS. Compounds analyzed were 1-Phthalic Anhydride, 2- Acetanilide, 3-Atropine, 4-IAEA C8, 5-Oxallic Acid II, 6-IAEA C6.*

Continuing work is optimizing operation with respect to the speed, samples size, software integration and modes of integration with the gas interface system.

[1] S. Fahrni et al., LIP annual report (2014) 26

[1] *Geology, ETH Zurich*

SIMPLIFIED RADIOCARBON MEASUREMENTS

A new direct gas interface system for radiocarbon analysis

M. Seiler, P. Gautschi, S. Fahrni, C. McIntyre, L. Wacker

The standard gas interface at ETH allows automated radiocarbon analysis of very small CO_2 samples [1]. However, the cleaning procedure of the zeolite trap and the syringe limits the sample throughput. These components were removed in a new design that is based on the direct flow of the sample gas into the ion source (Fig. 1). The system was built for the measurements of carbonates which can be decomposed in separate vials with phosphoric acid (85% H_3PO_4) to form CO_2 [2].

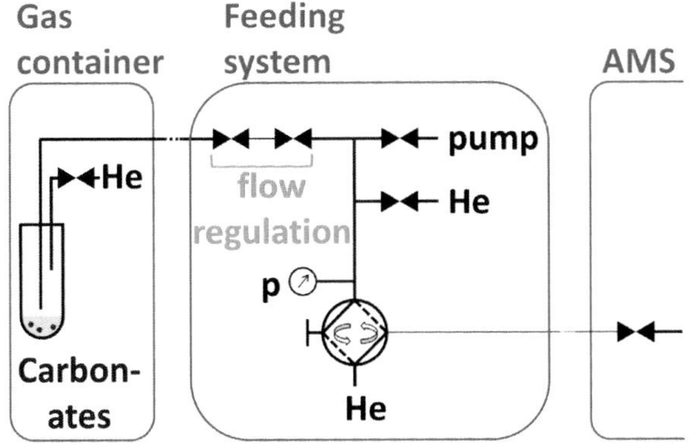

Fig. 1: *Schematic of the simplified gas feeding system for carbonate measurement.*

The vials are pressurized with helium to about 2 bar in order to achieve the desired concentration of 1-5% CO_2. The feeding system that connects the vial with the gas ion source, regulates the feeding pressure to maintain a constant carbon mass flow. This is achieved by limiting the inflow from the gas container by adjusting the opening times of the inlet valve.

A typical measurement starts with pre-sputtering the cathode followed by the initial adjustment of the feeding pressure (Fig. 2). This is needed because the CO_2 fraction in the gas is not exactly known. At the beginning of the measurement, the feeding pressure is kept constant by a computer program. This regulation works as long as the pressure in the vial is higher than the desired feeding pressure. To maintain high ion currents when the sample pressure gets too low, the vial is re-pressurized with helium. The additional helium dilutes the sample CO_2 but by increasing the feeding pressure the CO_2 flow is kept constant. With ongoing measurement time the pressure has to be increased more rapidly to compensate the dilution process (Fig. 2).

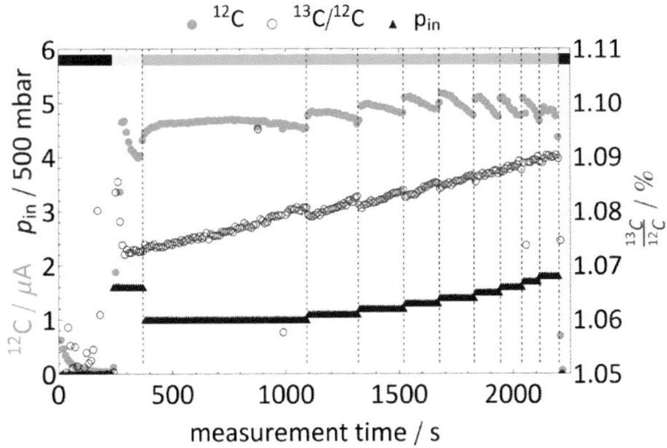

Fig. 2: *Key data from a radiocarbon measurement with the simplified gas interface as a function of time for a carbonate sample.*

For carbonates the system showed the best performance for samples in the mass range 50-100 µg C. For smaller samples the CO_2 concentration is too low and results in lower ion currents. This could be overcome by using smaller vials. The feeding system is not limited to measurements of carbonates only. A gas bottle filled with a CO_2–He mixture from combusted OX-2 was used for standard normalization. Also other gas containers are feasible, especially for samples that only have low precision requirements.

[1] L. Wacker et al., Nucl. Instr. & Meth. B 294 (2013) 315

[2] M. Schleicher et al., Radiocarbon 40 (1998) 85

EA-AMS COUPLING FOR ULTRAFAST ^{14}C ANALYSIS

Sample combustion and ^{14}C measurement in under 7 minutes

S.M. Fahrni, L. Wacker, C. McIntyre, H.-A. Synal

The coupling of an elemental analyzer to MICADAS (EA-AMS) has allowed direct sample combustion and measurement since 2010 [1]. However, a fully automated coupling for automated, unattended measurements has only been achieved recently.

There is an increasing demand for fast ^{14}C determinations with high throughput from fields such as biogeochemistry or biomedicine [2]. The fully automated EA-AMS coupling allows us to perform ultrafast combustion and analysis and also higher precision measurements of samples containing up to 100 µg C. Additionally, a stable isotope mass spectrometer for simultaneous measurement of precise $\delta^{13}C$ values (EA-IRMS-AMS) has been implemented for the first time [3].

Fig. 1: *The setup of the GIS (1) with coupled devices: carbonate handling system (2), elemental analyzer (3), and the stable isotope mass spectrometer (4). (Figure from annual report, 2013).*

A total analysis time of as little as 7 minutes per sample is achieved with a net data acquisition time of ca. 3 minutes (Fig. 2). The lower limit for the total analysis is given by the sample combustion and CO_2 separation in the elemental analyzer as well as heating and cooling of the zeolite trap (used for concentrating the CO_2 in the gas interface system [1]). The total analysis time could be further reduced to roughly 5 minutes by optimizing EA combustion settings, trap cooling and sample handling procedures.

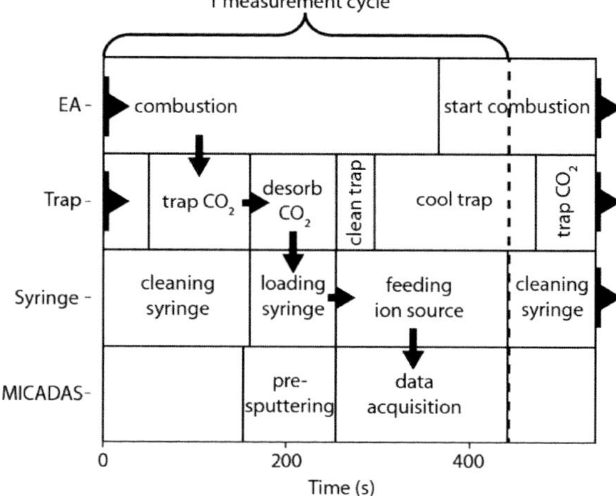

Fig. 2: *Typical timing of an EA-AMS run. The schematic depicts the process steps of the EA, trap, syringe and AMS in a continuous run. An entire measurement cycle is shown from 0 to 440 sec.*

EA-AMS measurements are mostly interesting for screening applications where fast and cheap radiocarbon analysis is desired. Higher precision measurements are feasible, too. However, the main limitations to the achievable precision are the sample mass and the capsule blank as well as sample cross talk, mostly due to the zeolite trap.

[1] M. Ruff et al., Radiocarbon, 52 (2010) 1645
[2] S. Fahrni et al., LIP annual report (2014) 27
[3] C. McIntyre et al., LIP annual report (2014) 24

BIOMEDICAL TRIAL ON MICADAS AND MyCADAS

First measurements of high throughput ^{14}C biomedical samples

S.M. Fahrni, L. Wacker, H.-A. Synal, S. Corless[1], M. Tucker[1]

A first set of low activity ($F^{14}C \leq 50$) biomedical samples was measured on the MICADAS and MyCADAS instruments as a continuation of our efforts in the field of biomedical AMS [1]. GlaxoSmithKline Research and Development provided the graphite and dried aqueous and plasma samples. Solid graphite samples were measured on both instruments, additionally dried liquid samples filled into tin capsules were measured as gas with the newly optimized EA-GIS-AMS coupling at MICADAS [2].

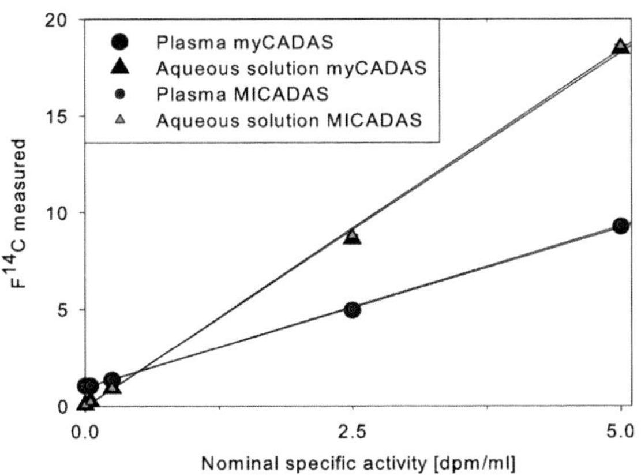

Fig. 1: Comparison of graphite samples from *dried aqueous solution and blood plasma.*

No significant differences were found between the well-established MICADAS instrument and the prototype MyCADAS (Fig. 1). Differences in $F^{14}C$ of samples with the same specific activities are due to different endogenous stable carbon amounts in plasma and aqueous solutions. Differences in diluting carbon in aqueous samples are also visible between solid (>1 mg C) and gas samples (<<1 mg C). Gas measurements, which took 7 minutes per sample, showed also a good reproducibility. However, it became apparent from the pronounced scattering of background values, that sample handling and dilution rather than AMS performance are the dominant source of error.

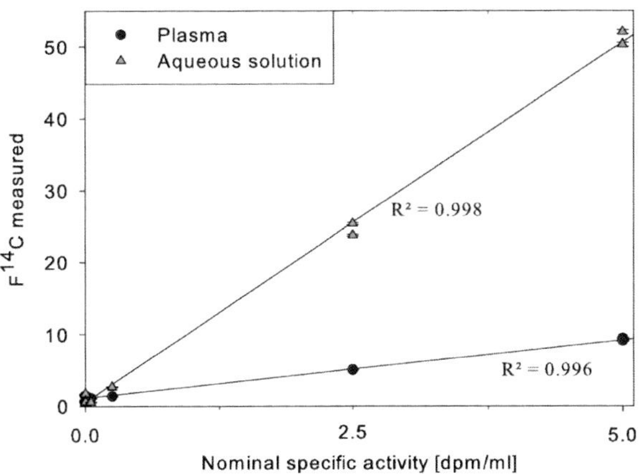

Fig. 2: Comparison of repetitive (n=2) gas *measurements. Values of aqueous solutions vary from solid measurements due to differences in sample preparation and dilution.*

With sample preparation being the most critical source of errors, the efficient and quick gas measurements are the obvious choice for such samples. Simple and fast sample handling and measurements will be key for future analyses of biomedical samples. To improve the robustness, future measurements should make use of the newly implemented EA-IRMS-AMS coupling [3]. It will be possible to employ precise $\delta^{13}C$ values as an internal standard for sample processing and dilution with carrier carbon. The "at line" approach with fraction collection has proven its suitability for such studies and we will continue our efforts in this field.

[1] T. Schulze-König et al. Nucl. Instr. & Meth. B 268 (2010) 891
[2] S. Fahrni et al., LIP Annual Report (2014) 26
[3] C. McIntyre et al., LIP Annual Report (2014) 24

[1] *GlaxoSmithKline R&D DMPK Ware, UK*

DETECTION AND ANALYSIS

Measuring ^{14}C ions below 50 keV

The upgrade of the PKU AMS - The final step

Digital pulse processing for RBS

MEASURING ^{14}C IONS BELOW 50 keV

Operating a simplified Bragg-type gas detector in proportional mode

A.M. Müller, M. Döbeli, M. Seiler, H.-A. Synal

The detection of single ^{14}C ions is very challenging at the energy range of less than 50 keV at which the so-called myCADAS AMS system operates [1]. Conventional mass spectrometers use electron multipliers (channeltrons, etc.) to detect single ions of a few keV. Disadvantages of this detector type are limited efficiency (<80%) and dark pulses, which are triggered without the impact of a primary ion.

In principle, gas ionization chambers (GIC) do not have these drawbacks. However, one 50 keV ^{14}C ion produces only less than 2200 charge carrier pairs during the stopping in common detector gases such as isobutane or argon. Conventional charge sensitive preamplifiers can hardly resolve such small signals. However, the signal can be amplified above the electronic noise level by operating the detector at higher voltages in the proportional region.

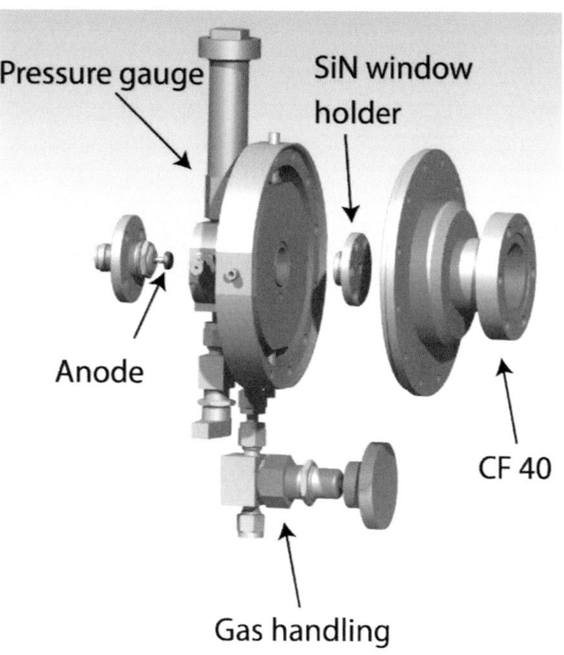

***Fig. 1:** View of the simplified Bragg GIC.*

First successful experiments at the myCADAS facility were performed with a simplified Bragg-type detector, which is routinely used as a ^{14}C counter at the 200 kV MICADAS system (Fig. 1). A thin 30 nm (4x4 mm^2) $Si_3N_{3.1}$ entrance window allows even very low energetic radiocarbon ions to enter the gas volume. Electron amplification in the gas was obtained by applying a 500 V bias voltage in 8 mbar isobutane.

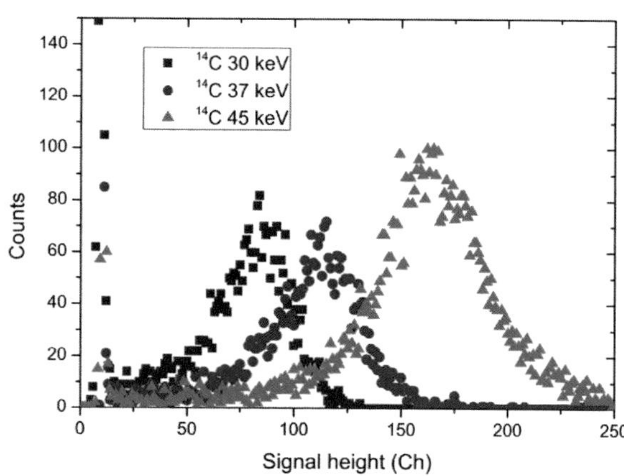

Fig. 2: *Energy spectra of radiocarbon measured at energies of 30 keV, 37 keV and 45 keV.*

Measurements were performed under these detector conditions for radiocarbon ions at 30 keV, 37 keV and 45 keV (Fig. 2) resulting in an energy resolution of 7-11 keV. A detection efficiency of more than 80% was determined by measuring ^{14}C standard samples. No dark counts could be observed. Because of these successful tests, the simplified Bragg-type detector is now installed permanently at the myCADAS as ^{14}C counter and more investigations on GIC in the proportional region will be performed.

[1] H.-A. Synal et al., Nucl. Instr. & Meth. B 294 (2013) 364

THE UPGRADE OF THE PKU AMS – THE FINAL STEP

First results of the extended PKU AMS facility

A.M. Müller, M. Christl, X. Ding[1], D. Fu[1], K. Liu[1], M. Suter, H.-A. Synal, L. Zhou[2]

At the beginning of 2014 an additional 90° bending magnet was installed at the PKU 500 kV AMS system (Fig. 1) after various preliminary studies during the past few years in collaboration with LIP [1]. The extension of the facility is based on an experimental setup implemented at the ETH Tandy system [2] and provides the possibility of measuring ^{10}Be with the degrader foil method. The radiocarbon measurement procedure and performance are not affected by the upgrade. The ^{14}C silicon detector is placed in front of the new magnet and can be retracted for the ^{10}Be measurement by a movable holder without breaking vacuum.

A gas ionization chamber (ΔE-E_{res}) provided by LIP was installed in the image plane of the new magnet in order to identify ^{10}Be ions. In March 2014 first experiments with ^{10}Be standard and blank samples were performed in order to investigate the performance. At 470 kV terminal voltage a ^{10}Be overall transmission (LE side into the detector) of 5-6% was achieved both for the 1+ and the 2+ charge state after the 75 nm SiN degrader foil.

Fig. 1: The extended beam line including the additional 90° magnet (350 mm radius) and the ETH gas ionization chamber.

The ^{10}Be/^{9}Be blank level for the 1+ charge state after the degrader was 2.4x10^{-15}, which is a factor of five lower than in 2+. It is not clear yet what causes this additional background.

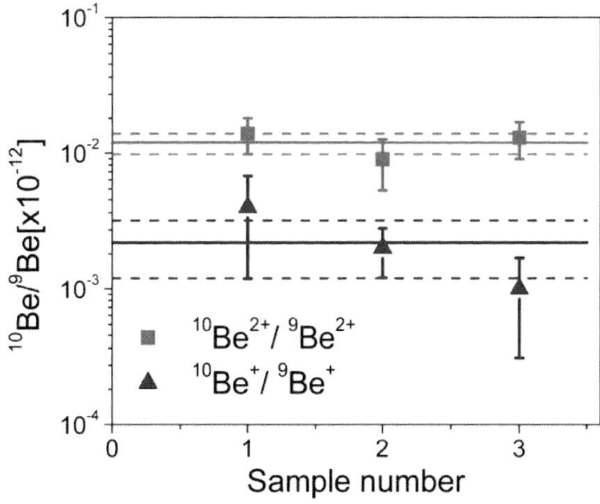

***Fig. 2:** The standard corrected ^{10}Be/^{9}Be blank ratios measured for the 1+ and 2+ charge state after the degrader.*

In conclusion, with this project it could be demonstrated that the dedicated radiocarbon NEC AMS facility CAMS can be upgraded with reasonable effort for ^{10}Be measurements and the obtainable performance is comparable with other compact AMS systems commercially available.

[1] A.M. Müller et al., Radiocarbon 55 (2013) 231

[2] A.M. Müller et al., Nucl. Instr. & Meth. B 266 (2008) 2207

[1] *State Key Laboratory of Nuclear Physics and Technology and Institute of Heavy Ion Physics, Peking University, Beijing, China*

[2] *Geography, Peking University, Beijing, China*

DIGITAL PULSE PROCESSING FOR RBS

IBA data acquisition system updated to latest technical standards

M. George, A.M. Müller, M. Döbeli

The traditional data acquisition (DAQ) chain of many ion beam analysis (IBA) systems, is based on analog equipment. This means that typically several devices with different signal processing functionalities are required for each data channel of an experiment. In contrast, recent developments in the field of digital pulse processing allow hosting all these functionalities in one device. Furthermore, each device can accommodate several readout channels. This reduces a four-channel DAQ to roughly 1/10 in size, as well as in cost.

For Rutherford Backscattering Spectrometry (RBS), we decided to develop a digital DAQ, based on the commercially available signal digitizer CAEN DT 5724. It provides on-board pulse processing, including pulse height analysis. The pulse height is determined by applying trapezoid fits to the recorded pulses, which follows the recommended approach for pulse height analyses [1].

The graphical user interface (GUI) (Fig. 1) was programmed in LabVIEW 2014. In the default acquisition mode, a standard configuration file containing all optimized fit parameters, can be loaded to run the program.

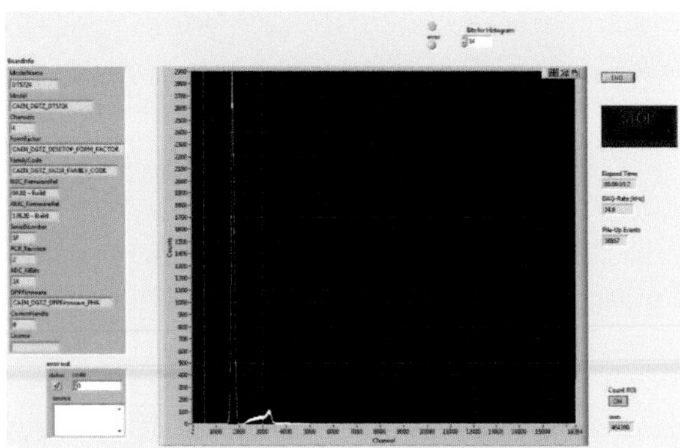

Fig. 1: *New GUI for RBS measurements.*

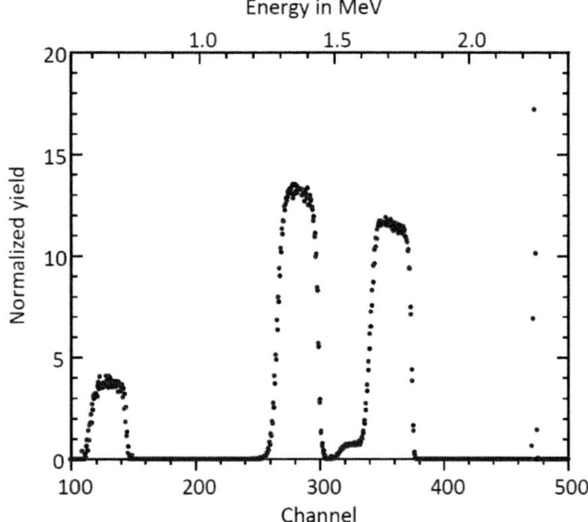

Fig. 2: *RBS spectrum of a barium titanate film. Overlay of data acquired with analog (black) and digital system (red).*

This default mode has been tested with a random pulser to rates of up to 100 kHz, which corresponds to the expected data rate for RBS times a safety factor of three. In an expert mode, the full pulse shapes can be displayed for parameter optimization or debugging.

In order to compare the performance of the existing analog DAQ chain to the newly developed fully digital DAQ chain, a series of samples has been measured with both systems. An example of overlaid spectra is shown in Fig. 2. There are no significant differences between the results. A pulser signal (sharp peak on the right of Fig. 2) proves that the resolution of the DAQ system is better than 8 keV and therefore is adding only a marginal contribution to the total energy resolution of the detection chain.

The same type of signal digitizer is planned to be used for further ion beam analysis experiments.

[1] V. Jordanov, G. Knoll, NIM A 345 (1994) 33

RADIOCARBON

Detail of: Bildnis Margrit mit roter Jacke und Konzertkleid, 1962; Artist: Franz Joseph Rederer; Copyright: Margrit Rederer, Zürich; Photo: SIK-ISEA, Zürich

The level of the Aegeri Lake in 1315 AD

Carbon burial on continental shelves

Stalagmite ^{14}C as a record for karst hydrology

The age and source of biomarkers in Lake Pavin

Utilization of permafrost in arctic freshwaters

Radiocarbon content of Swiss soils

Radiocarbon characteristics of kerogens

The Inn River drainage basin

The stratigraphy of margin sediments

Provenance of Fraser River organic carbon

^{14}C laboratory contamination testing

Rapid ^{14}C analysis of dissolved organic carbon

^{14}C LABORATORY IN 2014

Overview of samples prepared for ^{14}C analysis

I. Hajdas, G. Bonani, S. Fahrni, M. Maurer, C. McIntyre, M. Roldan Torres de Roettig, A. Synal, L. Wacker

The majority of samples submitted for ^{14}C analyses require pretreatment and conversion to graphite or CO_2 prior to AMS analysis. Activities in the ETH preparation laboratory can be summarized in terms of the number of samples that were prepared for different types of applications during the year 2014, as compared with previous years (Fig. 1, Tab. 1). This year was another highly successful year for the ^{14}C preparation laboratory.

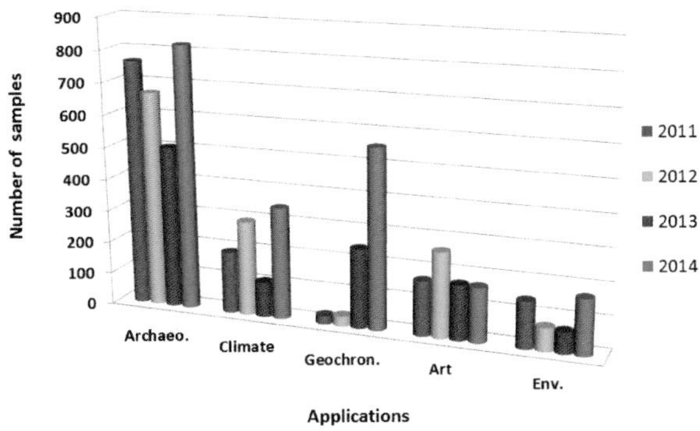

Fig. 1: *Number of samples (objects) analysed for various research disciplines during the last four years.*

Research	Total	Internal
Archaeology	821	16
Past Climate	347	11
Geochronology	557	204
Art	169	13
Environment	172	69
Total	**2066**	**313**

Tab. 1: *Number of samples analysed in 2014 for various applications. Column 'Internal' is the number of samples supported by the laboratory for master or term theses.*

With the exception of 'Art', the number of objects analyzed in 2014 increased for all applications. The increase in 'Geochronology' is due to research and master projects [1, 2]. The leading role of applications in 'Archeology' is the result of strong collaborations within Switzerland.

The numbers reported take into account the actual number of samples (objects) that were studied (Tab. 1). However, in many cases multiple AMS analyses were performed on various fractions. Indeed, the number of targets prepared by the laboratory was over 3300 if the unknown samples as well as reference materials are included. In addition 330 samples contained less than 200 μg carbon after pretreatment, were analyzed with the gas ion source (GIS). Among those, a large portion (73) of the targets was measured as a part of master thesis dedicated to dating paintings [3]. It is worth to note that most of the foraminifera samples were measured using GIS [4].

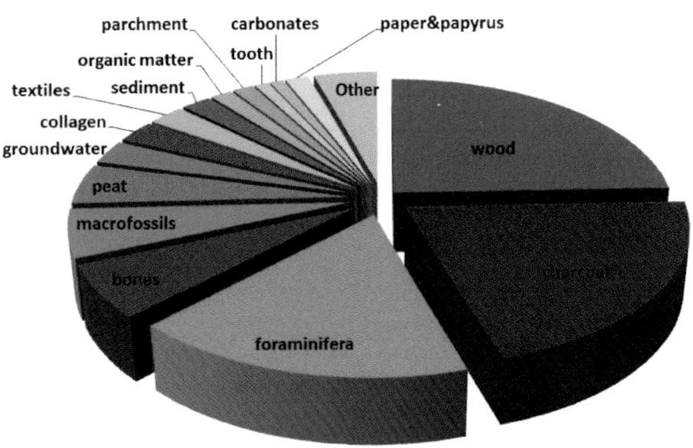

Fig. 2: *Type of material prepared and analysed at the ETH laboratory in 2014.*

[1] U. Sojc et al., LIP Ann. Report. (2014)
[2] K. Hippe et al., LIP Ann. Report. (2014)
[3] L. Hendriks et al., LIP Ann. Report. (2014)
[4] L. Wacker et al., LIP Ann. Report. (2014)

MICROSCALE RADIOCARBON DATING OF PAINTINGS

Application of ^{14}C dating as routine tool for artwork dating

L. Hendriks, I. Hajdas, E.S.B. Ferreira[1], M. Küffner[1]

Nowadays when the authenticity of a work of art is questioned, many aspects including material analysis, stylistic evaluation and historical context must be considered before reaching a final conclusion. Scientific analysis covers a broad range of methods from imaging techniques to non-destructive or minimal invasive analytical tools. Although material identification can assist in the determination of the oldest possible date of an artifact ^{14}C studies can provide reliable radiocarbon ages and have proven to be themselves decisive arguments in establishing forgeries [1]. However due to the relative large sampling requirement for ^{14}C dating, typically 10-30 mg of canvas [2], its application as routine dating method has been rather limited. Hence in collaboration with the Swiss Institute for Art Research (SIK-ISEA) the feasibility of using minimal sample sizes for ^{14}C dating as a routine tool was investigated.

Fig. 1: *"Brigitt" painted in 1962 by Franz Rederer (1899–1965). Calibrated ^{14}C ages of canvas predate the painting by 4-5 years.*

Prior to ^{14}C analysis the origin of the canvas fibre had first to be assessed, to exclude synthetic material that would afford misleading older ages. Canvas samples of unprecedented small size of only a few hundred µg were successfully dated using the gas ion source of MICADAS.

Fig. 2: *Pigment samples from painting by Karl Hosch (1900–1972). Trace amounts of synthetic organic dyes rendered inaccurate ^{14}C dating.*

Also the possibility of dating the organic binding medium was investigated. The characterization of the paint composition leading to the selection of suitable samples was carried out at the SIK-ISEA using X-ray fluorescence, Fourier transformed infrared spectroscopy and Raman spectroscopy. This sample scrutiny aimed at selecting paints free of carbon containing pigments and additives, which would be a source of error. The natural organic binders of paints were successfully dated to the time period of the painter's activity. These first results illustrate the potential of the developed method, hereby offering a new tool in the study of paintings and their origin, allowing detection of recent forgeries, copies and replicas.

[1] L. Caforio et al., Eur Phys J Plus 129 (2014) 1
[2] I. Hajdas et al., Radiocarbon 56 (2014) 637

---

[1] *Swiss Institute for Art Research, SIK-ISEA Zurich*

THE CHRONOLOGY OF HOLY SCRIPTURES

Writing surfaces from the Middle East dated – Parchment & Papyrus

M. Marx[1]*, E.-M. Youssef-Grob*[2]*, T. J. Jocham*[1]*, I. Hajdas*

The Middle East has a long history of written text, Holy Scriptures and the development of many different writing systems. The Corpus Coranicum project focusses on 3 main parts: 1. Textual evidence 2. Texts from the late antique environment of the Qur'ān and 3. A historical-critical commentary on the Qur'ān.

Research on the early qur'ānic manuscripts has shown that app. 1600-2000 *folium*. from the first centuries of Islam still exist in libraries and other collections. Unlike other documents however, those manuscripts are not bearing any date and need to be assigned a period in time by the shaping of the letterforms (palaeography) or orthography [1]. In a joint project between the DFG (Deutsche Forschungsgemeinschaft) and ANR (Agence National de la Recherche) material evidence is being dated to further the study of the qur'ānic text.

Fig. 1: *Sample extraction at the edge of the manuscript Ma VI 165 from the University Library of Tübingen.*

For establishing a broad set of data, almost 20 institutions have been approached for sampling of their precious manuscripts. Besides the parchment made of sheep skin (Fig. 1), dated papyri were included for comparative measurements, too. Their inclusion is not only important due to the fact that the first attempts to date qur'ānic manuscripts were based on them [2-3], but also to prove the validity of the ^{14}C dating method to the sometimes sceptic philologists.

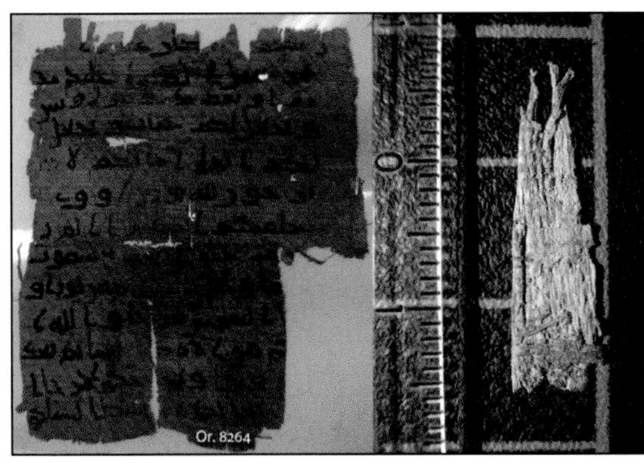

Fig. 2: *Rare qur'ānic fragment on papyrus from the University Library of Leiden.*

The first results of the dating campaign on early qur'ānic manuscripts – in a palaeographic style called ḥiǧāzī – are clearly indicating that those materials are dating back to the 7th century AD. With testing multiple targets and comparison between the different manuscripts, the radiocarbon analysis will yield very promising results that could counterbalance and complement the conventional dating methods.

[1] F. Déroche, Catalogue (1983)
[2] A. Grohmann, Der Islam 33/3 (1958) 213
[3] Arabische Paläographie (1967 & 1971)

[1] *Corpus Coranicum, Berlin-Brandenburg Academy of Sciences and Humanities, Germany*
[2] *Asien-Orient-Institut, University of Zurich*

ORIGINS OF PREHISTORIC ALPINE ANIMAL HUSBANDRY

Recent discoveries in the Silvretta range (Switzerland/Austria)

T. Reitmaier [1], I. Hajdas

Few regions in Europe are so strongly associated with Alpine animal husbandry and agriculture as the mountain regions of Switzerland and Austria. The seasonal use of high Alpine pastures by sheep, goat and cattle herds and the immediate and local utilisation of animal products seems perfectly adapted to the Alpine landscape, so much that this tradition continues into the 3rd millennium AD (Fig. 1). Perhaps this is part of the reason why origins and development of `Alpwirtschaft´ in the central Alps are still so badly understood. However, this might also be due to the methodological challenges of Alpine archaeology, and especially to difficulties in precise dating of the sites.

Fig. 1: *Rock-shelter with Neolithic occupation in Val Urschai, excavation during Sept. 2014 (Picture: Arch. Service of the Grisons).*

An interdisciplinary research project was initiated already in 2007 by the author (at that time working at the University of Zürich) to study the origins and development of Alpine animal husbandry in the Silvretta range on the Swiss-Austrian border. Starting points of the surveys were a number of settlements on the valley floor dating to the Bronze Age and Iron Age. Till 2014, during seven campaigns a large number of high Alpine (above 2000 m asl) sites dating between the earliest deglaciation and the modern age could be discovered. These included Mesolithic (e.g. ETH-39647, 9270 ± 45), Neolithic and Bronze Age abri sites as well as unique structures from the 1st millennium BC, such as animal pens (ETH-36464, 2210±45; ETH-36475, 1990±45) and huts (ETH-34341, 2425±55 BP). These are chronologically similar to and functionally complement the valley sites and are the very first architectural buildings precisely dated to the prehistory in the Swiss Alps [1].

The results of the still ongoing project show that the extensive Alpine pastures were being used from at least the 3rd millennium BC for summer grazing. These archaeological results are supported by e.g. archaeobotany/palynology, archaeozoology, toponymy and dendrochronology. The relatively high number of ^{14}C ages provides a fresh and detailed insight into the long-term history of human occupation and economy in the Swiss Alps [2-4].

[1] B. Dietre et al., Quat. Int. 353 (2014) 3
[2] T. Reitmaier et al., Archäologie Schweiz 36 (2013) 4
[3] Ph. Della Casa et al., Preistoria Alpina 47 (2013) 39
[4] T. Reitmaier (Ed.), Letzte Jäger, erste Hirten. Hochalpine Archäologie in der Silvretta (2012)

[1] *Archaeological Service of the Canton of the Grisons, Chur*

NEW LATEGLACIAL WOOD FINDINGS IN ZURICH

Can we extend the Radiocarbon Calibration Curve?

L. Wacker, U. Büntgen[1], B. Kromer[2], D. Nievergelt[1]

While still floating, the Lateglacial (LG) pines previously found in Zurich already played an important role in high-resolution studies of terrestrial ^{14}C calibration during deglaciation and the Younger Dryas [1]. Work on bridging the gap with already existing pines [2] came to a halt caused due to the death of Klaus Felix Kaiser in 2012.

In 2013 the prospect of closing the gap and extending the replication of the Swiss LG pine chronology became highly promising, when more than 250 new pines were found by WSL technician D. Nievergelt in the city of Zurich (*Binz material*). All of the new material was found about 200 m from the location where one of the earliest pines in the absolute master chronology from Kaiser were found in the 1980s. The stumps are generally well-preserved in silt and most can already be assigned to 14'000–11'000 BP based on 30 radiocarbon measurements performed on individual trees at the LIP combined with some dendro-chronological work done at the WSL.

Fig. 1: *More than 250 well-preserved stumps of pine trees were excavated in the Binz in Zurich.*

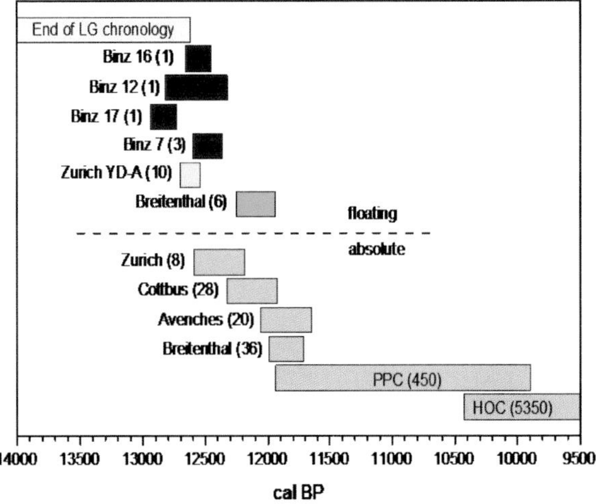

Fig. 2: *The important new (red and green) and old (blue) floating chronologies potentially extending the absolutely dated Holocene oak (HOC), Preboreal pine (PPC) and the Younger Dryas pine chronologies (gray). The number of trees per chronology is indicated in parenthesis.*

Figure 2 illustrates the end of the absolutely dated Preboral pine chronology with its extension to trees found in Zurich. Trees from the *Binz* (pre-dated with ^{14}C) may fill the gap to the end of the already existing floating Lateglacial chronology.

The presented work will be continued in the framework of a Swiss-German collaboration funded by SNF and DFG.

[1] R. Muscheler et al., Nature Geosci. 1 (2008) 263
[2] K.F. Kaiser et al., Quat. Sci. Ref. 36 (2012) 78

---

1 WSL, Birmensdorf
2 CEZA, Mannheim, Germany

CONFIRMATION OF TREE-RING DATING

Dendrochronological record confirmed with the 775 AD ^{14}C event

U. Büntgen[1], L. Wacker, K. Nicolussi[2], D. Güttler

Tree-ring-based temperature reconstructions represent the backbone of high-resolution paleoclimatology by providing useful long-term perspectives on global climate. A recent dispute regarding the potential misdating of ring-width chronologies due to 'missing rings', for trees growing near their thermal distribution limit, has raised questions about the reliability of tree-ring chronologies as annually resolved and absolutely dated climate proxy archives. The claim of missing tree rings, caused by exceptional summer cooling following large volcanic eruptions, is based on the experimental results of a cambial growth model, driven by simulated climate variations, which estimates tree growth during intervals of low temperatures [1]. So far, there has not been confirmation of the dating precision in temperature-sensitive ring-width chronologies using non-dendrochronological methods.

Fig. 1: *A subfossil pine from the Austrian Alps.*

To objectively investigate the so-called threshold-response hypothesis in tree-ring records, we consider a ring-width chronology from high-elevation settings in the Austrian Alps that continuously covers most of the Holocene [2]. Ten ^{14}C measurements were made on the dendrochronologically dated rings of a sub-fossil pine (*Pinus cembra*) excavated *in situ* from a peat bog at 2'100 m above sea level (Fig. 1).

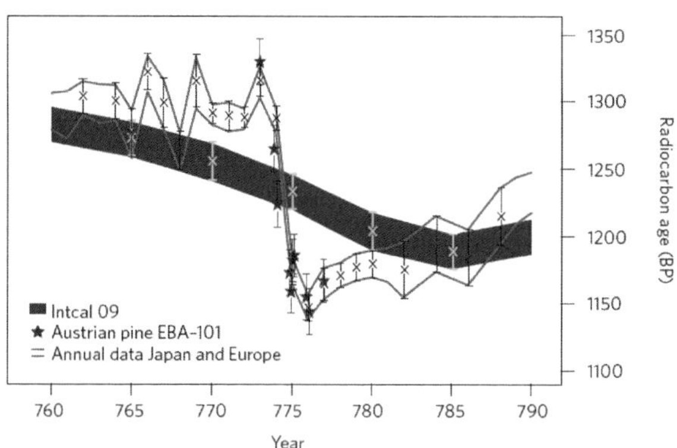

Fig. 2: *High-resolution radiocarbon ages (this study; red stars), superimposed on annually resolved ^{14}C measurements from Japan and Europe (grey lines and crosses) as well as the IntCal09 calibration curve based on decadal samples (blue shading).*

The measurements reveal a rapid 1.2% increase in ^{14}C concentration from the year 774 to the year 775, as well as a greater than 1.5% rise between 773 and 776 (Fig. 2). The observed peak coincides with high-resolution ^{14}C data (95% confidence interval) from Japan [3] and Germany [4]. Thus we conclude that the high-elevation record in the Austrian Alps is consistent with records from temperate sites in Japan, Germany and precisely date back to the year 775 AD. No single volcanic eruption was strong enough to trigger summer cooling sufficient to cause missing rings.

[1] M. Mann et al., Nature Geosci. 5 (2012) 202
[2] K. Nicolussi, et al, Holocene 19 (2009) 909
[3] F. Miyake et al., Nature 486 (2012) 240
[4] I.G. Usoskin et al., A&A 552 (2013) L3

[1] *WSL, Brimensdorf*
[2] *University of Innsbruck, Austria*

HIGH-RESOLUTION ^{14}C DATING OF DRILL CORE SEDIMENT

A 50-20 ka BP chronology from the Cormor alluvial megafan (NE Italy)

K. Hippe, A. Fontana[1], I. Hajdas, S. Ivy-Ochs

During the Last Glacial Maximum (LGM), the Cormor alluvial megafan was delivering large amounts of glacial sediment from the Alpine Tagliamento glacier onto the southern foreland basin. Rate and character of sedimentation were primarily controlled by glacier activity and, thus, by climate. To gain a better understanding of these processes, we have performed high-resolution radiocarbon dating of a 65 m long drilling core (PNC1, Fig 1) located in the distal sector of the Cormor alluvial megafan.

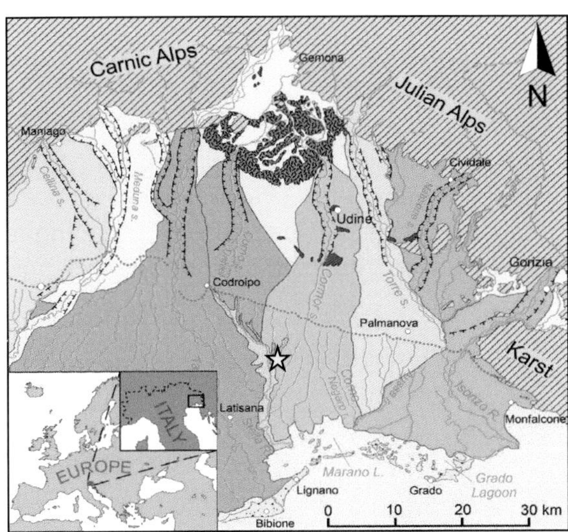

Fig. 1: *Alluvial megafans in the southern Alpine foreland (modified after [1]). The yellow star marks the coring site of PNC1.*

A series of 55 peat samples from 4-33 m depth were dated with radiocarbon. Obtained ages range from ~50-17 ^{14}C ka BP and provide a detailed chronology of the pre-LGM and LGM fluvioglacial sequence. 21 samples were separated into different fractions (Fig. 2) in order to check for systematic age differences with regard to the size and/or type of the organic particles. The influence of sample pre-treatment was evaluated by using the ABA treatment as well as two different ABOX treatment protocols for each sample fraction.

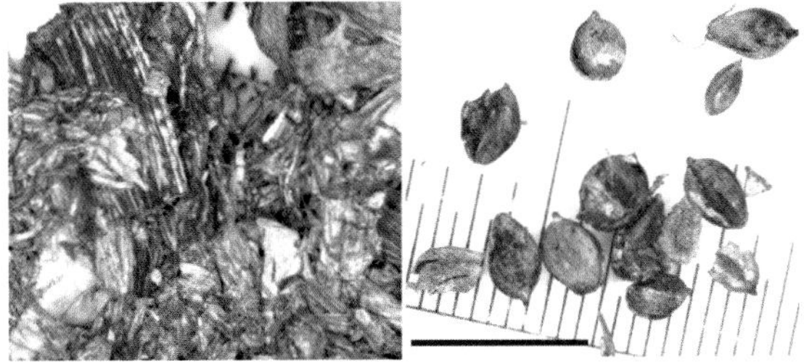

Fig. 2: *Dated bulk peat material (left) and separated seeds (right). Scale bar is 5 mm.*

While for samples at 4-10 m depth no systematic age difference is observed for the different sample fractions, results from below 10 m (older 19.5 ^{14}C ka BP) show significant variability (Fig. 3), which is subject to further investigation. Deposits containing old carbon (too old ages) were observed suggesting that a careful approach and high-resolution sampling is an imperative to obtain accurate chronologies.

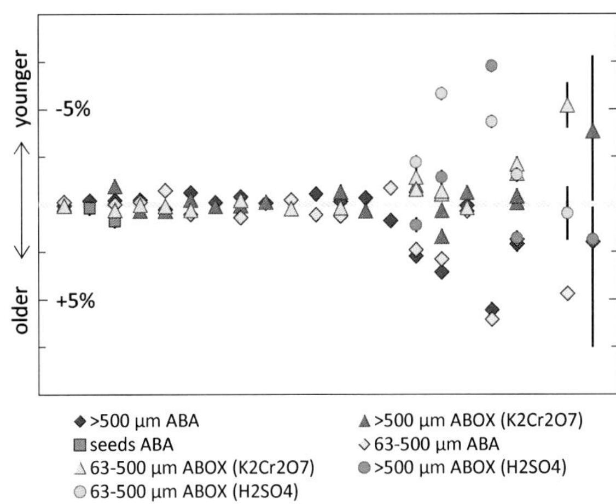

Fig. 3: *Variations in ^{14}C ages (1σ error) of sample fractions treated by ABA and ABOX.*

[1] A. Fontana et al., Geomorph. 204 (2014) 136

[1] *Department of Geoscience, Padua University, Italy*

RADIOCARBON DATING OF 50-40 KA OLD PEAT DEPOSITS

Reconstructing Middle Würm climate in the Swiss Alpine foreland

K. Hippe, I. Hajdas, S. Ivy-Ochs, M. Maisch[1]

The northern Alpine foreland plays a key role in the investigation of Quaternary climate evolution and the reconstruction of the glacial history of the Alps. However, the extent and timing of the various phases of glacier advance and retreat is still debated, in particular for the time prior to the Last Glacial Maximum (LGM).

To improve the understanding of the phase of ice build-up during the Middle Würm (~50 to 25 ka), we have performed radiocarbon dating of compressed peat deposits ("Schieferkohle") in the Swiss Alpine foreland (Fig. 1). These deposits of up to several meters thickness are truncated by LGM glaciofluvial and glacial sediments and are one important archive for late Pleistocene climate changes. The presence of peat is interpreted to correspond to phases of glacier-hostile, interstadial climate conditions.

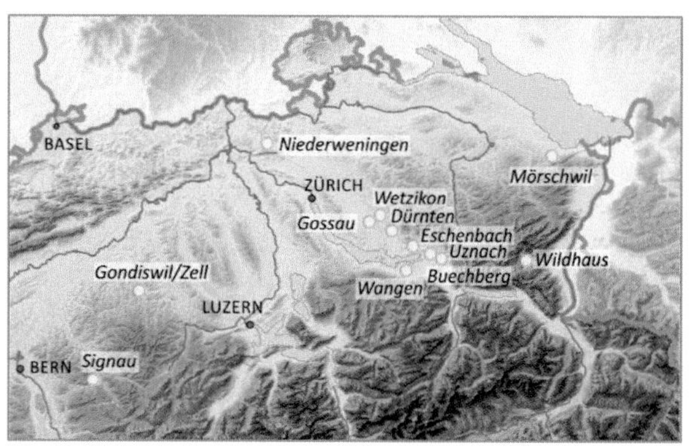

Fig. 1: *Major deposits of compressed peat in the Swiss Alpine foreland.*

Around the town of Dürnten, SE of Zürich, compressed peat has been mined since the 19th century. Today, peat deposits can still be found along stream cuts or at construction sites. Based on detailed pollen analysis in drill cores from Dürnten, Welten [1] reconstructed stadial-interstadial cycles since the end of the Riss glaciation and throughout the Würm. On one of these well-preserved drill cores (Fig. 2), we were able to obtain radiocarbon ages ranging from >50 to 43 ka BP. These allow to place the palynological data into an absolute timeframe and to establish a detailed chronology of the Middle Würm climate variations.

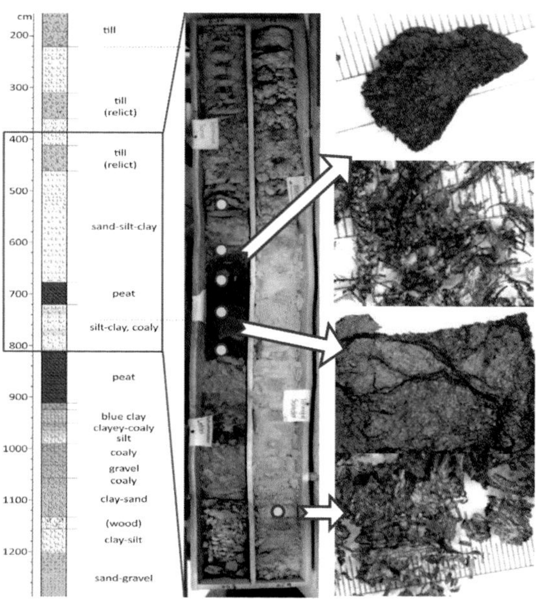

Fig. 2: *Upper part of the investigated drill core with sampling spots for radiocarbon dating. Pictures on the right show different material separated for dating (bulk peat, macrofossils).*

^{14}C ages of 50 to 45 ka BP obtained from surface outcrops are in excellent agreement with the core data as well as data from other peat deposits along the valley [2], [3].

[1] M. Welten, Beiträge zur Geologischen Karte der Schweiz, N.F. 156 (1982) 174 pp

[2] C. Schlüchter, Vierteljahresschrift Naturf. Ges. Zürich 132 (1987) 135

[3] I. Hajdas et al., Quat. Int. 164-165 (2007) 98

[1] *Geography, University of Zürich*

THE TIMING OF THE LGM IN NORTH-EASTERN ITALY

Outlooks from radiocarbon dating of alluvial deposits

A. Fontana[1], K. Hippe, L. Hendriks, S. Ivy-Ochs, I. Hajdas

Despite the huge footprint left by the Last Glacial Maximum (LGM) glaciers in the Alpine valleys and morainic amphitheatres, the availability of organic material for radiocarbon dating is scarce and a detailed geochronology of the glacial phases that occurred during the LGM is still poorly documented. A particular situation characterizes the Venetian-Friulian Plain (NE Italy), where alluvial megafans were built during the LGM and their distal sector is dominated by fine deposits with the rather common presence of peaty and organic horizons [1]. These megafans were directly fed by the fluvioglacial supply of the Alpine glaciers. High sedimentation rates are partly connected to the glacial advances and pausing of glaciers at their front. We studied two stratigraphic cores collected along the mainland boundary of the Grado-Marano Lagoon (Fig. 1), where the top of LGM surface is extensively exposed and henceforth it is possible to study these deposits also through a geological and geomorphological survey (Fig. 2).

Fig. 1: *Oblique aerial picture of the area of Stella River mouth and location of the PNC core.*

In the Piancada core (PNC, Fig. 1) around 55 samples were collected between 4 and 33 m depth and radiocarbon dated with the aim of testing different pre-dating methods for materials of the LGM and pre-LGM age. The data demonstrated that the LGM fluvioglacial deposition started at almost 26.5 m, which was around 30 ka cal BP, and ended 10 ka later forming the present surface. An average depositional rate of 2.7 mm/yr can be calculated but lower and higher values can be observed for specific intervals, probably in relation to different fluvioglacial phases.

Fig. 2: *Stratigraphic section of alluvial deposits at Piancada (PNC) dating back to the end of the LGM.*

In the core of Beligna collected near the Roman city of Aquileia, 12 samples were dated. The first part of the LGM is probably not recorded, but the fluvioglacial deposits at Belinga started from 28.6 m of depth and reached the present surface where dune-like features were observed. The investigation demonstrates that these surface landforms were formed ca. 21 ka ago, towards the end of the LGM, when fluvioglacial activity ended.

[1] A. Fontana et al., Sed. Geol. 301 (2014) 150

[1] *Geosciences, University of Padua, Italy*

ROOTS AND RADIOCARBON DATING CHRONOLOGIES

Dating late Holocene landslide event in the Mont Blanc area, Italy

U. Sojc, I. Hajdas, S. Ivy-Ochs, N. Akçar[1]*, P. Deline*[2]

The Ferret valley Arp Nouva peat bog located in the Mont Blanc massif was critically evaluated since early ^{14}C ages published previously [1] are not coherent with recently obtained ^{10}Be ages [2] leading to controversial conclusions on the formation of the swamp. Radiocarbon dating of organic matter from three pits of up to 1 m depth was applied to discuss the question whether the historical documented rock avalanche occurring in AD 1717 overran the peat bog or formed it at a later stage.

Fig. 1: *Profile of peat bog Arp Nouva 2 sampled for ^{14}C dating.*

Samples of peat (Fig. 1) were wet sieved so that woody fragments of roots could be identified (Fig. 2). In most radiocarbon chronologies, roots penetrating the deposits are a potential contaminant in dating bulk sediments. Specifically fine roots, being almost invisible, pose problems to such chronologies. In our study, roots were not modern, which would be expected from roots penetrating the sediments, indicating that the trees were killed in the event. Indeed, radiocarbon ages of all the roots were in agreement with the historically documented AD 1717 rock avalanche event. It can therefore be concluded that the rock avalanche formed the Arp Nouva peat bog by downstream blockage of the Bellecombe torrent. Furthermore, careful sample preparation with consequent separation of roots from the bulk peat sample has shown that the problem of too old ^{14}C ages can be circumvented.

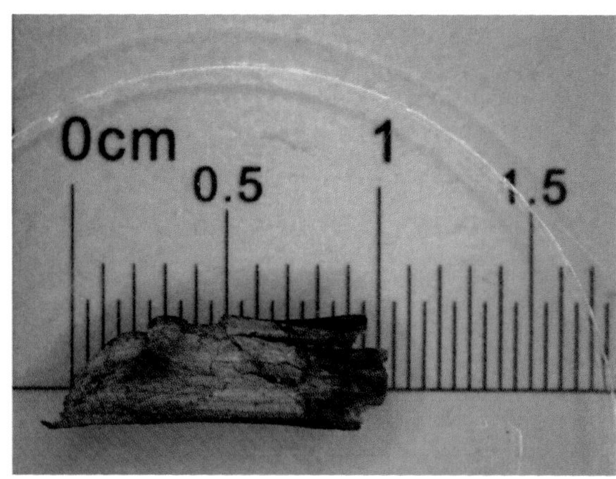

Fig. 2: *Roots of trees killed in the AD 1717 rock avalanche.*

This work demonstrates that a combined geomorphological and geochronological approach is the most reliable way to reconstruct landscape evolution, especially in light of apparent chronological problems. The key to successful ^{14}C dating is careful sample selection and identification of material that might be not ideal for chronological reconstructions.

[1] P. Deline and M. Kirkbride, Geomorph., 103 (2009) 80

[2] N. Akçar et al., J. of Quat. Sci., 27 (2012) 383

---

[1] *Geological Sciences, University of Bern*

[2] *EDYTEM Lab, Université de Savoie, France*

TREE-RING ^{14}C ANALYSIS

Volcanic CO_2 for tracing eruptions on Mount Etna (Sicily, Italy)

R. Seiler[1,3], P.Cherubini[1], N. Houlié[2], F. Moergeli[3], I. Hajdas

Early indicators of volcanic activity would be instrumental in hazard management but are still lacking. An increased vegetation index (Normalised Difference Vegetation Index) showing photosynthesis activity was measured from satellite imagery along an eruptive fissure on Mount Etna (Sicily, Italy) [1]. Additionally, we found higher growth-rates of trees growing along two eruptive fissures prior to eruptions, confirming the described higher photosynthesis rates. Tree growth, as a result of tree-crown photosynthesis processes, is an expression of environmental factors and predominantly depends on water availability, temperature, air CO_2 concentrations and soil nutrients. Tree growth in volcanic areas seemingly is affected by volcanic below-ground processes (Fig. 1). Volcanic degassing of mainly H_2O, CO_2, SO_2 or H_2S, usually precedes eruptions on Mount Etna. To understand which volcanic processes are triggering tree-growth enhancement, wood chemical analyses need to be carried out.

Fig. 1: *Trees growing right next to lava flows on Mount Etna (Photo. R. Seiler).*

Since trees largely depend on CO_2 for their photosynthesis we hypothesised that volcanic CO_2 might have caused the observed growth increase before the eruption. We measured ^{14}C in annual tree rings formed before and after the 2002-2003 flank eruption, to assess tree CO_2-uptake prior to eruptions. However, our preliminary results reveal no tree-ring ^{14}C depleted values, showing that trees did not assimilate any significant volcanic ^{14}C - dead CO_2 (Fig. 2).

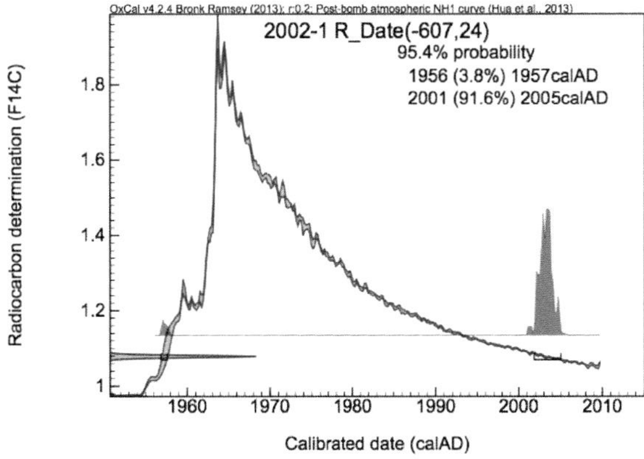

Fig. 2: *Results of ^{14}C analyses of tree rings growing before the 2002/3 flank eruption show no depletion in ^{14}C.*

In summary, thanks to the ^{14}C analyses we can exclude emissions of magmatic CO_2 as a trigger of increased tree rings growth, which helps to narrow the spectrum of factors influencing the growth of the trees.

[1] Houlié et al., 246 (2006) 231

---

[1] *WSL, Birmensdorf*
[2] *Geodesy and Geodynamics Lab, ETHZ, Zurich*
[3] *Geography, University of Zurich*

MASS MOVEMENTS AT 700 AD IN LAKE SILS, ENGADINE

Further age constraints for a lake-wide catastrophic mass movement

R. Grischott[1], F. Donau[1], F. Kober[2], I. Hajdas, M. Strasser[1]

Previously radiocarbon-dated sediment cores from the Lagrev basin of the Lake Sils in Upper Engadine (Fig. 1) indicated an age for a catastrophic delta slope failure of 700 AD. There was little known of what happened in the westward situated Maloja basin [1]. It was argued that this mass movement was triggered by an earthquake. Because earthquakes are able to trigger multiple slope failures across the entire lake [2], there is a need to study the westward situated Maloja basin for mass movement deposits.

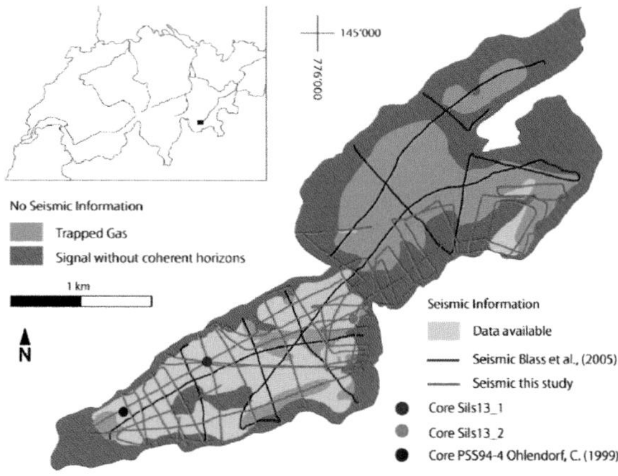

Fig. 1: Lake Sils with locations of sediment cores and seismic survey from this study and previous work [1]. Note the Lagrev basin (right upper corner) and the Maloja basin (left lower corner).

In September 2013, a seismic survey of the western and main part of Lake Sils was carried out in order to define the optimal coring site. The retrieved sediment cores were opened, lithologically described, and sampled for age datings. A few needles marked as sample ETH-55744 (Fig 2.) yielded ^{14}C age that date the event close to 700 AD.

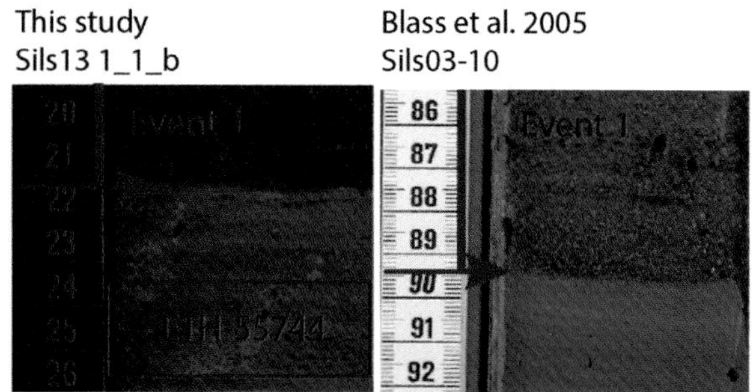

Fig. 2: The sample was taken 2 cm below the base of the event deposit thus giving a maximum age estimate. Note the same deposit previously observed by [1] in the Lagrev basin (right).

Given the age range of [1] with 650-780 AD and 530-670 AD, our sample is closer to the younger age for the catastrophic mass movement. As the existence of these deposits can also be observed in the shallower Maloja basin, one can state that there was an event which was occurring across the entire lake at around 700 AD. The most likely trigger for this event is suggested to be external like an earthquake, probably induced by activity related to the Engadine fault line.

[1] A. Blass, et al., Ecl. geol. Helv 98 (2005) 319
[2] M. Strasser et al., Geology 34 (2006) 1005

[1] *Geology, ETH Zurich*
[2] *NAGRA, Wettingen*

YD AGE FOR THE UNIQUE STAPPITZ LAKE ARCHIVE (A)

Age dating of sediment from a former sparsely vegetated environment

R. Grischott[1], F. Kober[2], J. Reitner[3], I. Hajdas, S. Ivy-Ochs and S. Willett[1]

Previously radiocarbon-dated sediment cores from the Stappitz lake archive in Carinthia, Austria (Fig. 1) were limited to the Early Holocene and a core depth of 55 m [1]. Since the entire core is 160 m long, it was speculated whether the core extends back to the Lateglacial.

Fig. 1: *Lake Stappitz from the West with view to the glacially shaped Seebach-Valley. The core location is located on the currently inactive delta behind the boat building.*

The lithology of the sediment core is rather fine grained with the largest fraction between silt to fine sand with sparse organic matter. An additional problem resulted in the conditions in which the core which was stored after the drilling campaign in 1996. The entire core dried out and sediment layering was no longer visible. As picking organic samples by hand was not possible, we had to tediously sieve the sediment by half meter steps starting at a depth of 55 m and an age of around 10600 cal BP. At a depth of 66.5-67 m a piece of wood was found and dated to the age that lies within the Younger Dryas Period (e.g. [2]).

Given the Younger Dryas age in 66.5 m depth, one can easily assume that the deeper core parts must be of Lateglacial age. Considering the fact that the lake sediments in the core reach down to -118 m one can hypothesize these sediments possibly range back to the Bölling-Alleröd Interstadial. In the context of Quaternary evolution, we can interpret the depositional environment as a proglacial basin filled in by an oscillating valley glacier during the Lateglacial.

[1] A. Fritz and F.H. Ucik (Sonderband Nationalpark Hohe Tauern (2001)

[2] S. Ivy-Ochs et al., Ecl. Geol. Helv. 89 (1996) 1049

[1] *Geology, ETH Zurich*
[2] *NAGRA, Wettingen*
[3] *Geology Bundesanstalt, Vienna, Austria*

RATES OF DEADWOOD DECAY IN AN ALPINE FOREST

A combined approach using ^{14}C and tree-ring analyses

M. Petrillo[1], M. Egli[1], P. Cherubini[2], H.-A. Synal

Coarse woody debris (CWD), i.e. deadwood, has a well-documented role in forest functioning: it provides habitat for many autotrophs and heterotrophs and is a preferential site for tree seedlings [1]. Due to the highly heterogeneous spatial distribution of CWD and its long-term decay dynamics, only little quantitative data about the CWD decay dynamics exist. A five decay-class system is usually applied to describe the decay stage of CWD. This system is based on a visual assessment of CWD in the field. Quantitative data with respect to the age of CWD in these decay classes is in most cases entirely missing. We investigated the CWD decay rates in Alpine forests dominated by Norway spruce (*Picea abies* (L.) Karst) and European larch (*Larix decidua* Mill.) in Val di Sole and Val di Rabbi (Trentino; Italy). Wood cores from living trees and cross sections from CWD were taken at the selected sites. At each site, 5 or 6 living trees were sampled to build a reference (master) ring-width chronology for each species, i.e., Norway spruce and European larch. All CWD samples were first classified into the decay classes. Tree-ring width was measured, crossdated and analysed in the dendrochronological laboratories at WSL, Birmensdorf. The correlations among all the ring-width series of living trees and CWD were statistically assessed using the software COFECHA. CWD of the decay classes 4 and 5 was too strongly degraded to be dated by using tree ring-analysis. The wood structure in these decay classes was too altered and tree rings no longer visible. In such cases, the outermost part of the CWD was sampled and ^{14}C-dated (using the AMS facilities at ETHZ). The average age of the CWD did not differ among the decay classes 1 to 3. Although these decay stages are visually distinctly different, the rate at which decay processes occur seems to vary to such an extent that a differentiation in the age is not possible. The ages for larch and spruce CWD were in the range of 3–40 and 1–54 years, respectively.

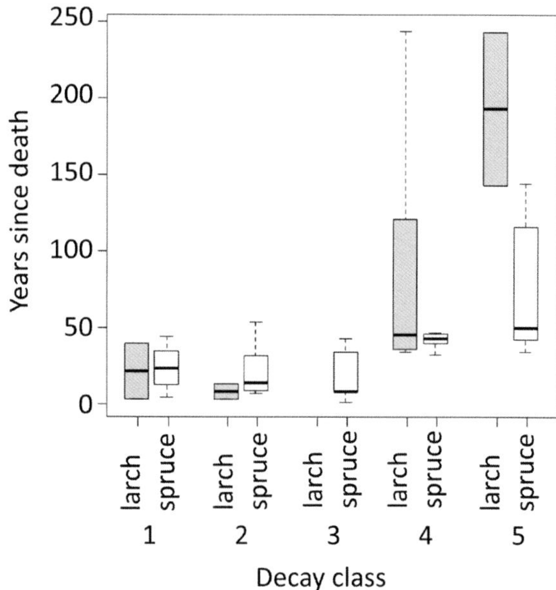

Fig. 1: *CWD age as a function of tree species and increasing decay stage (decay class).*

In decay class 4 and 5 distinct tree species-specific differences were found. Larch CWD reaches, as an average, an age of 200 years before the wood residues are fully integrated into soil organic matter (this process is faster for spruce, Fig. 1). With our analyses we were able to derive quantitative estimates and empirical decay constants for larch and spruce. Depending on the tree species, mean residence time of carbon can be similar to soil organic matter.

[1] J. Rondeux and C. Sanchez, Environ. Monit. Assess. 164 (2010) 617

---

[1] *Geography, University of Zurich*
[2] *WSL, Birmensdorf*

THE LEVEL OF THE AEGERI LAKE IN 1315 AD

A multi-methodological approach to trace landscape evolution

M. Egli[1], *M. Maisch*[1]

In the year 1315, the Habsburgs fought a battle at Morgarten (Central Switzerland) against the Swiss Confederation. Although important for the Swiss foundation history, this battle is more and more controversially discussed. Many historical facts are missing which gave rise to speculations and the designation of the battle as the 'myth of Morgarten'. The number of involved warriors and the exact location of this battle is still a matter of debate. The outcome of this battle seemed to be driven by the landscape in general and the lake's dimension (Aegeri lake) in particular.

Two hypotheses about the lake level and landscape in 1315 AD persisted so far: i) the lake level was about 6 m higher and additional lakes or pools existed in side valleys versus ii) the lake's dimension was not different and the situation was similar to today (i.e. 724 m asl).

A multi-methodological approach was used to precisely reconstruct the evolution of the lake's dimensions and the landscape. This included geomorphic mapping of former lake terraces, chemical and physical characterisation of mires close to the lake of Aegeri, radiocarbon dating of the peat basis, geoelectric exploration of the sediments, analysis of existing soil maps [1] and evaluation of existing archaeological data. Based on this procedure a model of landscape evolution was derived. In the Late Glacial, the lake level was much higher than nowadays (at about 750 – 760 m asl). Due to a catastrophic outburst event, the lake level decreased by ca. 25 m. About 5500 BP, the lake level was at 732 m asl and during the Roman period at 724 – 726 m asl. Early medieval graves mark a maximum lake level at 724 or 725 m asl in the period 500 – 1050 AD.

At the time of the battle at Morgarten, the lake level was most likely at an altitude of 726 m to max. 727 m asl and consequently about 2 – 3 m higher than today. This is documented by ^{14}C dating, a lake terrace at this altitude and by the spatial extension of the most developed soils.

Fig. 1: *Age of peat bases near Morgarten*

Core drillings in the surroundings of the lake and in neighbouring hollows showed that additional pools existed at that time. Due to the cooler climate in 1315 AD, the valley floors were moister and the fens more widespread. During the 15th to the 17th century, part of these pools disappeared. Due to a more intense agriculture (draining of fens), the fen area shrunk.

Regarding the lake level, hypothesis ii) seems to better reflect the situation for 1315 AD. However, small lakes and pools in adjacent depressions were confirmed as given by hypothesis i). Although landscape scenery and lake levels can now be drawn quite precisely for this period, it still remains unclear where exactly the battle took place. According to old paintings, the adversarial warriors were attacked with stones, boulders (pushed from steep slopes) and tree trunks. Many who tried to escape drowned in a nearby lake. But there where the lake is, the slopes are not steep and there where steep slopes prevail only small pools seemed to exist. Consequently, the myth of the battle of Morgarten still lives on.

[1] M. Egli and P. Fitze, Catena 46 (2001) 35

---

[1] *Geography, University of Zurich*

CARBON BURIAL ON CONTINENTAL SHELVES

Quantification in deltaic and non-deltaic sediments

R. Bao[1], C. McIntyre, M. Zhao[2,3], T. Eglinton[1]

The transfer and burial of terrestrial organic carbon (OC) in continental marginal seas is a critical source-to-sink process in carbon cycling. Large amounts of terrestrial OC is returned to the atmosphere before and after sedimentation due to remineralization [1], however a significant proportion is not recycled and accumulates in depocenters on continental shelves. The accumulation of OC in the sediments can be linked to supplies of material from different carbon reservoirs such as OC mobilized from the terrestrial and marine biosphere (recent OC, OC_{recent}) [2], or eroded from sedimentary rocks (petrogenic OC, OC_{petro}) [3]. Petrogenic OC is more refractory and resistant to remineralization processes allowing a greater proportion to be sequestrated in the sediments and influence the marine carbon cycle [1]. Although many recent studies concern the fate and preservation of terrestrial OC from surface sediments on continental shelves, the actual percentage that escapes degradation during the sedimentation has remained elusive.

We collected sediment cores from two sites in the East China Sea (ECS) representing a typical subaqueous muddy prodeltaic zone (P01) and an alongshore mobile muddy non-deltaic zone (ME3). We measured the ^{14}C Fm and total organic carbon content (TOC) down each core to determine the proportions of OC_{petro}. According to Galy et al., (2008), a plot of TOC versus modern OC % (Fm x TOC) gives a regression line with an X-axis intercept that is equal to the percentage of OC_{petro} preserved in a sample (Fig. 1). The cores P01 and ME3 had OC_{petro} contents of 0.03 % and 0.1 %, respectively. These results show that the proportions of OC_{petro} in the deltaic area is lower than that in the non-deltaic areas.

The slope of line is equivalent to the Fm of the OC_{recent} in a core and constrains the percentage of OC_{petro} that is finally buried. The slopes for P01 and ME3 are 0.70 and 0.85, respectively (Fig. 1). Pre-aging and loss of OC can occur though suspension in hydrodynamic loops so it is consistent that the Fm of the OC_{recent} is reduced in the highly energetic deltaic area at P01. The data presented here suggest that the slope (Fm of the OC_{recent}) can be used in estimating the percentage of OC_{petro} (intercept of x-axis) burial. This finding, that the fate of OC_{petro} is constrained by OC_{recent} adds to our understanding of the process of OC_{petro} burial in the river-dominated continental shelves.

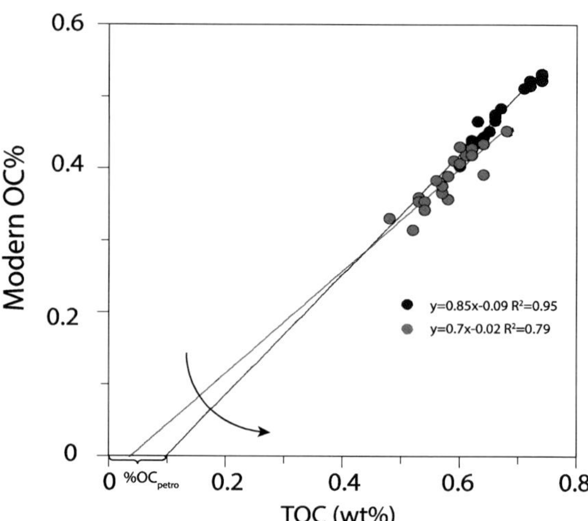

Fig. 1: *TOC versus Modern OC for the sediment cores P01 (green) and ME3 (red) from the East China Sea. The slope and x intercept are used to calculate the percentage of petrogenic OC.*

[1] B. Mckee et al., Cont. Shelf Res. 24 (2004) 899

[2] V. Galy et al., Science 322 (2008) 943

[3] R.G. Hilton et al., Nat. Geo. 1 (2008) 759

[1] *Geology, ETH Zurich*

[2] *Marine Organic Geochemistry, Ocean University of China, Qingdao, China*

[3] *Marine Chemistry Theory and Technology, Ocean University of China, Qingdao, China*

STALAGMITE ^{14}C AS A RECORD FOR KARST HYDROLOGY

Towards reliable ^{14}C-dating of speleothems

F. Lechleitner[1], C. McIntyre, T. Eglinton[1], J. Baldini[2]

Stalagmites are a widely used archive for paleoclimatic and environmental reconstructions. Their popularity stems from their ubiquitous distribution over all continents, the possibility of multi-proxy studies on a single sample, and very precise absolute dating using the U-Th method.

Fig. 1: *Drilling of carbonate samples from a small stalagmite slab. A semi-automatic drill is used (Sherline 5400 Deluxe).*

However, a number of speleothems are not amenable to U-Th dating, leading to the abandonment of potentially important cave sites and samples in the past. This problem could be circumvented by using ^{14}C to date speleothems, if the "dead carbon" problem is solved. Carbonate bedrock dissolution adds a variable amount of ^{14}C-dead carbon to the dripwater solution that feeds the stalagmite, resulting in a reservoir effect called the dead carbon fraction (DCF).

DCF has been proven to be variable over time, and to respond very sensitively to changes in karst hydrology, often related to climate changes [1,2]. It can therefore be used as a stalagmite proxy by itself, as we show in a high-resolution stalagmite DCF-record from the Yok Balum cave, Belize (Fig. 2).

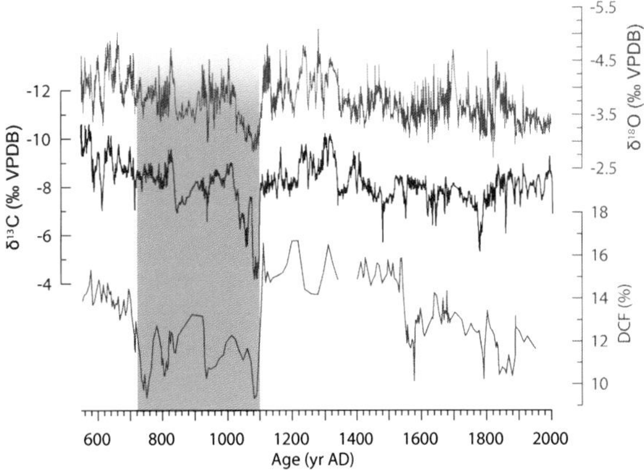

Fig. 2: *Stable oxygen and carbon isotope and DCF records from stalagmite YOK-I from Yok Balum cave, Belize. The blue bar indicates a period of prolonged drought in Belize, which is reflected in changes in the isotopes and DCF.*

Using this relationship with hydrological changes, an attempt has also been made to model DCF. By using other speleothem proxies that reflect karst processes and hydrology, we are able to reproduce general trends in DCF and larger shifts related to hydrology.

For this purpose, we processed more than 300 speleothem carbonate samples in 2014, resulting in unprecedented spatial and temporal resolution of the records (up to 0.1mm/sample).

[1] M.L. Griffiths et al., Quat. Geochron. 14 (2012) 81

[2] A.L. Noronha et al., EPSL 394 (2014) 20

[1] *Geology, ETH Zurich*

[2] *Earth Sciences, Durham University, UK*

THE AGE AND SOURCE OF BIOMARKERS IN LAKE PAVIN

^{14}C measurement of specific sedimentary *n*-alkanes and alkanoic acids

M. Makou[1], T. Eglinton[2], C. McIntyre, D. Montluçon[2], V. Grossi[1]

Compound-specific radiocarbon measurements have provided important source and transport constraints for organic carbon (OC) preserved in marine environments, but similar information is lacking for lacustrine settings. The increasing use of organic geochemical proxies to generate paleoclimate records from lakes requires further investigation of biomarker fidelity. We measured the radiocarbon content of specific *n*-alkane and alkanoic acid homologues preserved in surficial sediments from Lake Pavin in order to better understand their biological sources and the relative ages of these common proxy compounds.

Fig. 1: *Location of Lake Pavin in France.*

Lake Pavin is a meromictic volcanic crater lake (maar) located on the Massif Central in southern France (Fig. 1). With excellent sedimentary OC preservation over the last 7,000 years and a strong water column dissolved inorganic carbon ^{14}C gradient [1], it is an excellent setting for investigating lacustrine OC cycling. Several sampling campaigns (Fig. 2), a previous bulk ^{14}C study [1], and the present ^{14}C investigation of sediments, lake particulate matter, and adjacent soils have provided the material necessary for molecular source attribution and interpretation of reservoir residence times.

Fig. 2: *Water column sampling campaign. Steep crater walls with mixed vegetation provide a continuous source of terrestrially derived OC.*

The *n*-alkanes and alkanoic acids found in sedimentary environments span a wide range of carbon chain lengths, with the longest homologues commonly attributed to vascular plant leaf waxes and the shortest to aquatic microorganismal sources. The old (millennia) compound-specific radiocarbon ages obtained here suggest the this relationship is substantially more complex in Lake Pavin, with a much wider range of microorganismally-derived homologues than is commonly observed or interpreted. In contrast to other investigations and common assumptions, long-chain alkyl compounds derived exclusively from higher plants with rapid transport to the sediments were not in evidence. These results suggest that proxies involving these compounds should be employed with caution in lake settings.

[1] Albéric et al., Radiocarbon 55 (2013) 1029

---

[1] *Laboratoire de Géologie de Lyon, Université Claude Bernard Lyon 1, France*
[2] *Geology, ETH Zurich*

UTILIZATION OF PERMAFROST IN ARCTIC FRESHWATERS

Age (^{14}C) of fluvial organic carbon utilized by Arctic microbes

P. Mann [1], R. Spencer[2], N. Zimov[3], C. McIntyre, T.Eglinton[4]

On-going climate warming in the Arctic is leading to extensive thawing of previously frozen permafrost soils. Northern circumpolar terrestrial permafrost contains vast quantities of organic carbon (OC), comprising as much as 50% of global below ground soil carbon stocks [1]. Once thawed, permafrost OC becomes available for microbial decomposition, leading to the release of greenhouse gases (CO_2 & CH_4). A proportion of permafrost OC is also mobilized to extensive freshwater systems, prevalent across Arctic landscapes. Once in freshwaters systems, the fate and bioavailability of permafrost OC however is largely unknown.

To examine the age and microbial availability of OC in Arctic waters, we measured ^{14}C and ^{13}C of dissolved OC (DOC) from fluvial networks across the Kolyma River Basin (Siberia), and isotopic changes during incubation experiments (Fig. 1).

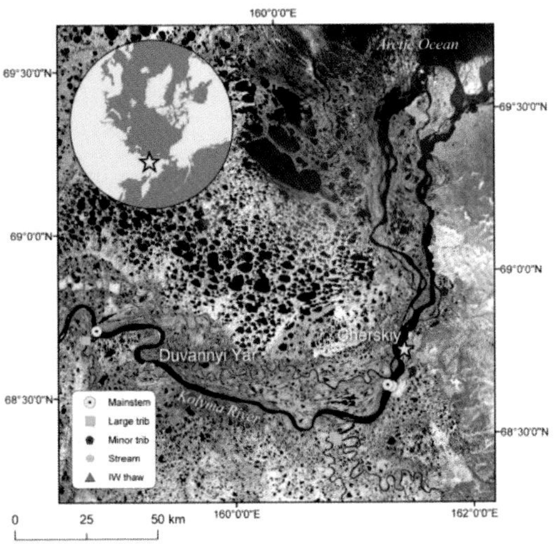

Fig. 1: *Sampling sites across Kolyma River Basin*

Microbes in waters directly draining permafrost outcrops consistently used DOC highly depleted in ^{14}C (-750 to -1000 ‰) indicating the biological loss of Pleistocene-aged DOC. The $\Delta^{14}C$ values of DOC utilized by microorganisms downstream varied considerably ($\Delta^{14}C$ +300 to -630 ‰) indicating a disconnect between bulk DOC age and the age of DOC utilized by microbes. Mean $\Delta^{14}C$-DOC utilized by microorganisms became higher (younger) moving downstream, with an apparent convergence upon a modern OC in major tributary waters and main-stem sites.

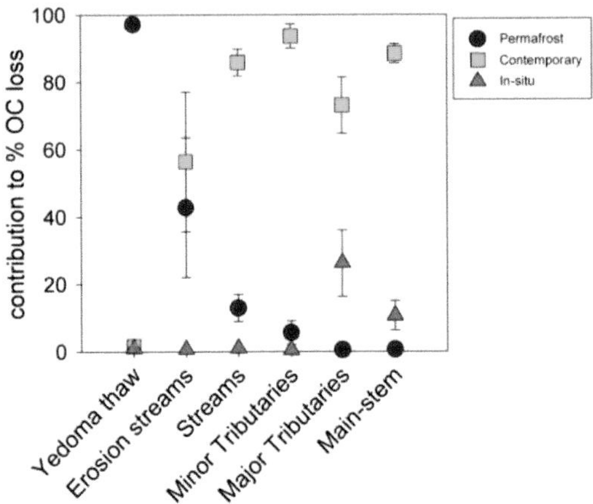

Fig. 2: *Contribution of permafrost, in-situ derived DOC and contemporary surface soil carbon to microbial demand across the basin.*

Using a dual-isotope approach (Fig. 2), we estimate that microbial demand in thaw waters is subsidized by up to 97 % by permafrost DOC. Downstream in large rivers and main-stem waters < 1 % of microbial turnover is derived from permafrost DOC. Our results indicate a preferential loss of permafrost DOC occurs during fluvial inland water transit.

[1] C. Tarnocai et al., Global Biogeochem. Cycles. 23, (2009) GB2023

---

[1] Geography, Northumbria University, UK
[2] Ocean and Atmospheric Science, Florida State University, USA
[3] Pacific Institute for Geography, Far-Eastern Branch of Russian Academy of Science, Cherskiy, Russia
[4] Geology, ETH Zurich

RADIOCARBON CONTENT OF SWISS SOILS

Scales of spatial and temporal variability across different regions

T. van der Voort[1], C. Zell[1], C. McIntyre, X. Feng[2], F. Hagedorn[3], P. Schleppi[3], T. Eglinton[1]

Soil organic matter (SOM) forms the largest terrestrial pool of carbon outside of sedimentary rocks, and it provides the fundamental reservoir for nutrients that sustains vegetation and associated microbial communities. Radiocarbon is a powerful tool for assessing OM dynamics and is increasingly used in studies of carbon turnover in soils. However, due to the nature of the measurement, comprehensive ^{14}C studies of soils systems remain relatively rare.

***Fig. 1:** The Beatenberg site, high spatial variability end-member site.*

In particular, information on spatial variability in the radiocarbon contents of soils is limited, yet this information is crucial for establishing the range of baseline properties and for detecting potential modifications to the SOM pool, as well as setting boundaries for large-scale carbon stock modeling. This study aims to develop and apply a comprehensive four-dimensional approach to explore heterogeneity in bulk SOM ^{14}C, with a broader goal of assessing controls on organic matter stability and vulnerability in soils across Switzerland. Focusing on range of Swiss soil types, we examine spatial variability in ^{14}C as well as ^{13}C, C % and C:N ratios over plot (decimeter to meter) to regional scales, vertical variability from surface to deeper soil horizons. The highest variability end-member site (Beatenberg) is shown in Figure 1. Sites are part of the Long-term Ecosystem monitoring program of the WSL.

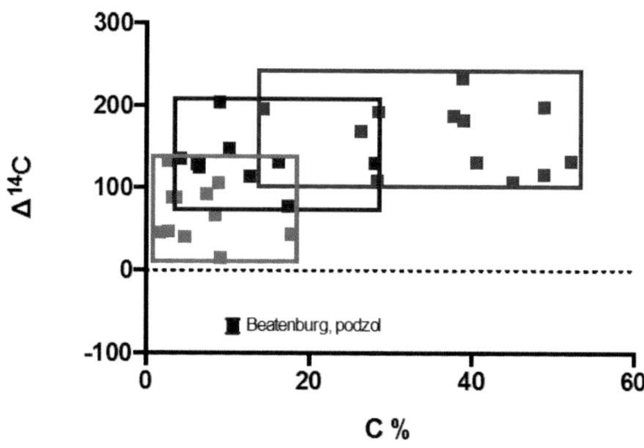

***Fig. 2:** Variability in radiocarbon signatures of sites spanning climatic and geologic regions in Switzerland.*

In 2014 we performed more than 150 radiocarbon measurements and results show that differences in SOM ^{14}C age across small lateral and vertical distances within soil systems (10^1-10^2 m, Figure 1) can be as large as those between regions (10^4-10^5 m, Figure 2). Results will be followed up by spatial analysis on different scales as well as compound-specific analyses and eventually modeling studies.

---

[1] *Geological Institute, ETH Zurich*
[2] *Chinese Academy of Sciences, Beijing, China*
[3] *Biogeochemistry, WSL, Birmensdorf*

RADIOCARBON CHARACTERISTICS OF KEROGENS

Releasing covalently-bound components via hydropyrolysis

M. Sieber[1], C. McIntyre, W. Meredith[2], T. I. Eglinton[1]

Catalytic hydropyrolysis (hypy) has proven to be a reliable technique to release covalently bound biomarkers from kerogens while keeping their carbon skeletons, stereochemistry, and isotopic signatures intact (Love et.al 1995, Meredith et al. 2010). Here, we apply hypy to sediments from four different depositional settings: Arabian Sea (marine upwelling region), Bermuda Rise (sediment drift site), Lake Constance (lacustrine with strong riverine input) and Mackenzie River (river bank sediment) in order to release covalently-bound biomarkers for ^{13}C and ^{14}C carbon isotope analysis.

Carbon isotope analysis of kerogens and their hypy fractions revealed that organic carbon in the kerogen fraction is older than in the solvent-soluble bitumen fraction suggesting that covalently-bound compounds are conserved within the kerogen over longer timescales (Fig. 1), protected from degradation by the surrounding macromolecular structure.

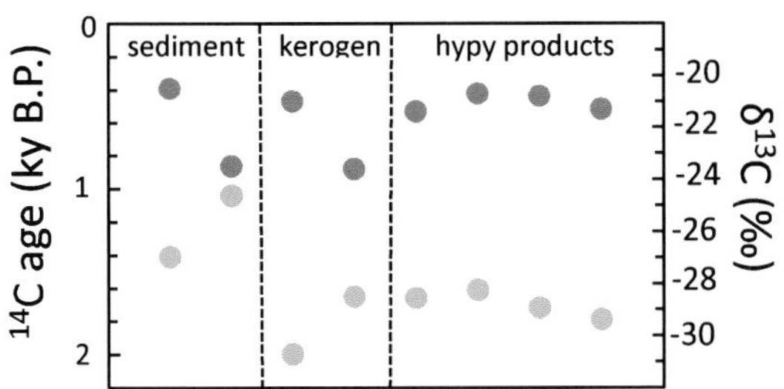

Fig.1: *$\delta^{13}C$ (blue) and ^{14}C age (green) for sediment (total organic carbon and total lipid extract), kerogen (total organic carbon and total lipid extract) and hypy products (bulk and three temperature windows) (from left to right) of the Arabian Sea sediment.*

Hypy products show consistent carbon isotope ratios for each site and in combination with previous data, we were able to infer sources of OC in the kerogen. The main hydrocarbons released via hypy were identified as n-alkanes, methylalkanes and aromatic hydrocarbons. We were able to isolate specific biomarkers from the hypy products by preparative chromatography as well as measure their radiocarbon ages by accelerator mass spectrometry (Fig. 2).

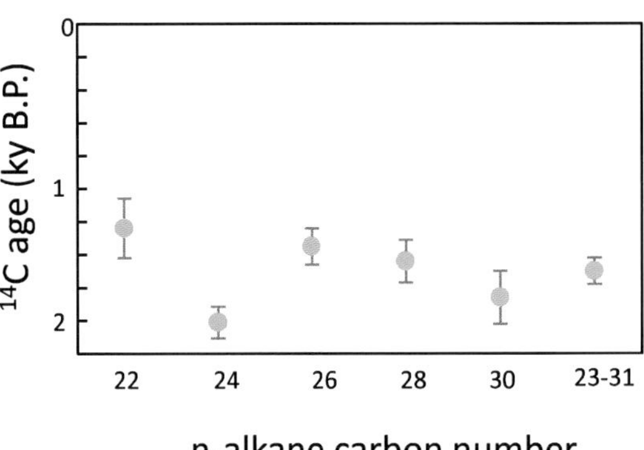

Fig.2: *^{14}C age for the isolated n-alkanes (numbers denote carbon chain length) from the Arabian Sea sediment using hypy.*

In conjunction with the bulk data for the hypy fractions, our results are in agreement with the findings of previous studies that hypy is a suitable treatment to release covalently-bound biomarkers for compound-specific isotope analysis and that ^{14}C data can reveal detailed age structure information in the subfractions and molecules of sedimentary material.

[1] G. Love et al., Org. Geochem. 23 (1995), 981

[2] W. Meredith et al., Rapid Commun. Mass Spectrom. 24 (2010) 501

---

[1] *Geological Institute, ETH Zurich*

[2] *Engineering, University of Nottingham, UK*

THE INN RIVER DRAINAGE BASIN

Composition and provenance of terrestrial biomarkers

N. Kündig[1], A. Gilli[1], B. Buggle[1], C. McIntyre, T. Eglinton[1]

Organic matter exported by rivers to continental and marine sedimentary archives is an important process in the global carbon cycle. To provide more insights into the provenance and transportation history of organic matter, we studied the Inn River system (Fig. 1). This system is one of the largest rivers that drains the eastern part of the Alps and it has a complex terrain and climate with steep slopes, different lithology and vegetation types. Riverbank sediments and suspended particulate matter (SPM) samples were collected along the river and at its main tributaries and analyzed for a variety of parameters. These included bulk carbon composition, ^{13}C and ^{14}C carbon isotopes and lipid biomarkers.

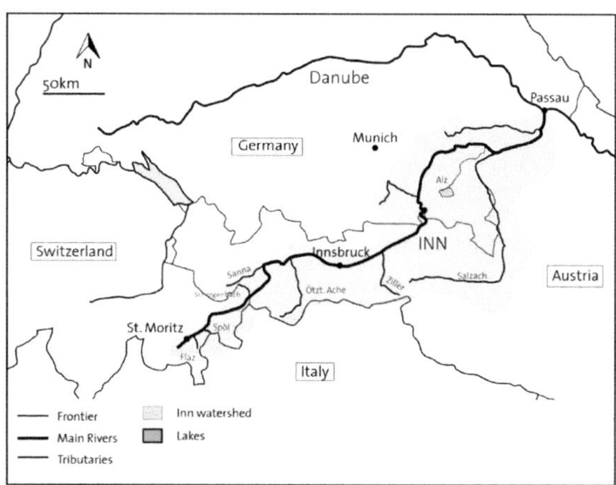

Fig. 1: *The Inn River Basin.*

Total organic carbon (TOC) values along the Inn River ranged from 0.16-2.78 wt% with low organic carbon loadings (OC/SA). The bulk ^{13}C ranged between -23.56 to -29.55‰ which is usual for a C3 vegetation cover. A coupled molecular isotopic mass balance approach based on bulk ^{13}C and ^{14}C Fm data was used to make a quantitative assessment of sources (rock, soil, aquatic) of organic carbon in the samples. Results showed that soils accounted for 40-80% of the organic carbon, 20-40% is from fossil OC and only 0-20% comes from aquatic in-river production (Fig. 2).

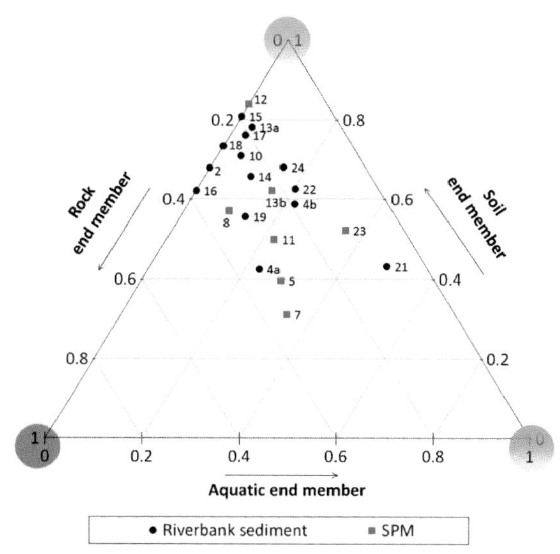

Fig. 2: *Provenance analysis of organic matter in the Inn River basin.*

The branched isoprenoid tetraether (BIT) index which indicates soil inputs was high and further supported these results. The distribution of n-alkanes and n-alkanoic acids in the Inn River samples suggests that fresh vascular plants are a main source of the organic carbon, with more grass input in the upper basin evolving into a tree dominated input downstream. Generally the results showed higher variations in the upper Inn basin due to a greater diversity of local inputs. These developed into a more constant, integrated signal downstream as individual sources were mixed.

---

[1] Geology, ETH Zürich

THE STRATIGRAPHY OF MARGIN SEDIMENTS

Investigating sediment transport using radiocarbon dating

P. Wenk[1], *C. McIntyre, C. Magill*[1], *T. Eglinton*[1]

Marine sediments serve as an important archive for palaeoclimate research. A broad variety of proxies which can be applied to sediments exist for the identification and interpretation of past climate changes and it is crucial that biomarker signals can be precisely dated. Diverging biomarker ages within a single sediment layer can occur due to the varying origins and transport paths of the deposited matter. Here we dated organic carbon from separated grain size fractions from a kasten core from the Iberian margin to better understand these processes. This is an area known to have high sedimentation rates and one that expresses a 'bipolar seesaw' signal due to contrary climate behaviors of the N and S hemispheres during the late Pleistocene.

The studied interval covers the period since the last glacial termination. Grain size fractions were chosen based on hydrological properties. The core was sliced into layers of 1 cm. The sediment was grain size fractionated by sieving down to 63 μm, further centrifuged to get the clay fraction (<2 μm) and finally settled to get a separation from the coarse and the fine silt at 10 μm (Fig. 1).

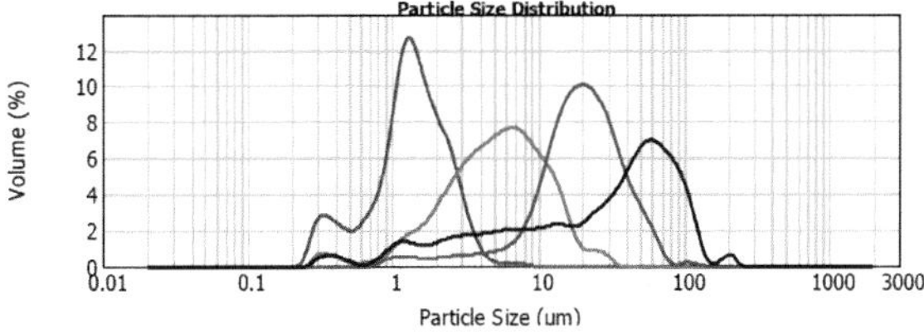

***Fig. 1:** Check of grain size fractionation using a Malvern Laser Diffraction Grain Sizer. red: clay <2 μm, green: fine silt 2-10 μm, blue: coarse silt 10-63 μm, dark blue: sand 63-200 μm*

The resulting record exhibits ^{14}C age differences within a single sediment layer of more than 5000 years (Fig. 2). The age differences exhibit a grain size-specific dependency as well as varying amplitudes. Preliminary interpretation suggests that both of these aspects are a result of changes in the palaeocurrent speed. The largest age amplitudes correlate with times of expected fast flow speeds [1]. Furthermore, the age of the fine and the coarse silt fractions also depend on the current speed. During supposed high current speeds the coarse fraction exhibits the oldest age while during slow speeds both show similar ages. Potentially, this could be explained by changing redistribution and lateral transport processes that affect the separated grain size fractions in different ways.

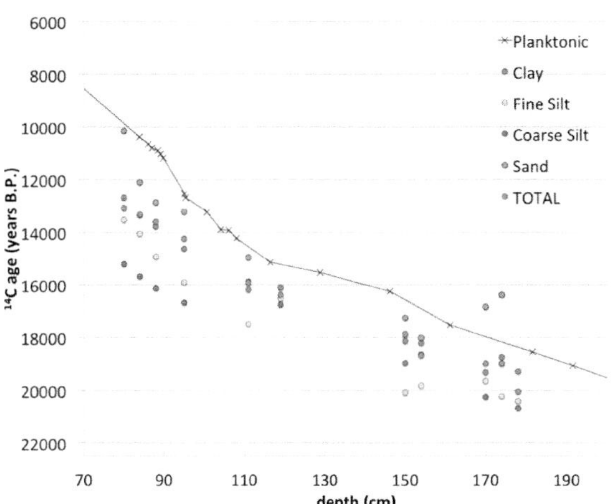

***Fig. 2:** Organic carbon ^{14}C ages of grain size fractions from various layers within an Iberian Margin sediment core. The ages diverge strongly inside individual sediment layers (colored points). Foraminiferal ^{14}C data (crosses) is from a nearby core and indicates that the sedimentation rate correlates negatively with flow speed in this region.*

[1] I.R. Hall and N. McCave, Earth Planet. Sci. Lett. 178 (2000) 1

---

[1] *Geology, ETH Zürich*

PROVENANCE OF FRASER RIVER ORGANIC CARBON

Probing sources of organic matter using ^{14}C

B. Voss[1], B. Peucker-Ehrenbrink[1], T. Eglinton[2], C. McIntyre

The Fraser River in southwestern Canada drains a mountainous area of diverse climate, vegetation, and geology. A long-term study of the geochemical composition of dissolved and sedimentary material carried by the Fraser River has revealed significant spatial heterogeneity among tributary basins and seasonal cycles in dissolved inorganic geochemical tracers [1]. The current work aims to characterize the age composition of organic matter across the basin. Spatial variability in climate, terrain, and vegetation type are expected to impart distinct signatures to organic matter originating in different areas.

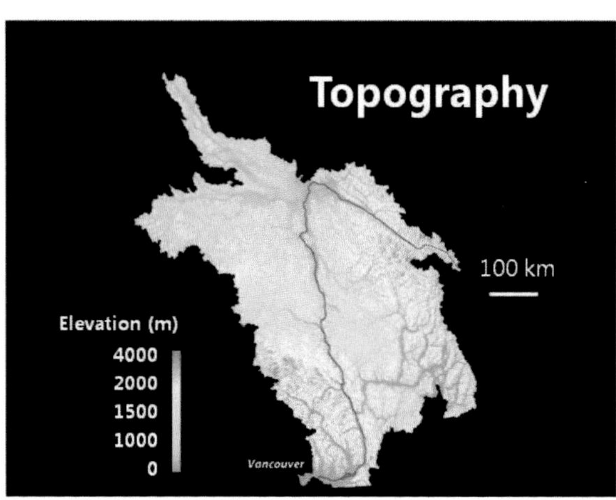

Fig. 1: A relief map of the Fraser basin indicates zones of mountainous terrain and deeply incised canyons, which influence river sediment mobilization.

Beginning in 2009, water and sediment samples were collected from tributary and main stem sites across the basin under low, medium, and high discharge conditions. Various types of samples were prepared for ^{14}C analysis. Suspended sediment samples were fumigated in acid vapor to eliminate mineral carbonates, and bulk organic carbon was extracted and isolated as CO_2 on a vacuum line. Sediments were also solvent extracted to purify biomarkers of terrestrial higher plants for ^{14}C analysis.

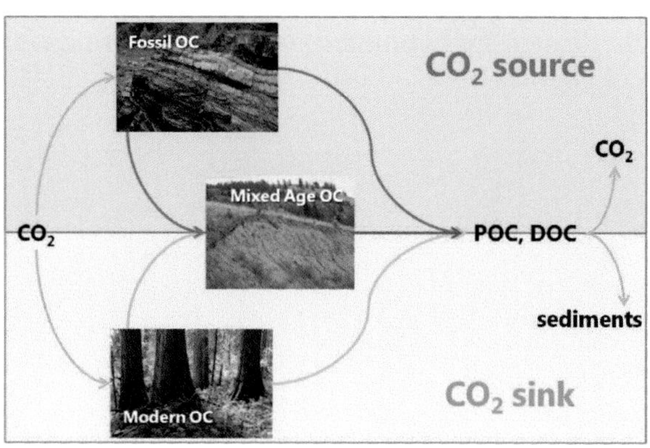

Fig. 2: Organic carbon carried by river sediments influences long-term climate variations through burial of recently fixed atmospheric CO_2. ^{14}C age allows for the quantification of relative inputs of aged versus modern sources of sedimentary organic matter.

Bulk organic carbon ages for Fraser River sediments were found to range from greater than modern to >5800 ^{14}C years. On average, Fraser sediments were found to contain ~0.12 wt. % of fossil organic carbon. Long-chain fatty acid biomarkers extracted from river sediments were older than bulk sedimentary organic matter from the same sample, with ages ranging from 1950 ($C_{24:0}$) to 9200 ^{14}C years ($C_{32:0}$). Such old soil biomarker ages indicate ongoing release of pre-aged soil organic matter following the deglaciation of this region over the past 10-12 kyr.

[1] Voss et al., Geochim. Cosmochim. Acta, 124 (2014) 283

[1] *Marine Chemistry & Geochemistry, Woods Hole Oceanographic Institution, Woods Hole, USA*
[2] *Geology, ETH Zurich*

^{14}C LABORATORY CONTAMINATION TESTING

Rapid screening using a wet chemical oxidation method

C. McIntyre, F. Lechleitner[1], S. Lang[2], L. Wacker, S. Fahrni, T. Eglinton[1]

The use of radiochemicals enriched with ^{14}C can contaminate work areas used for natural abundance radiocarbon measurements. It is often difficult to know whether equipment, samples or facilities have been affected so, prior knowledge and testing is invaluable to ensure the isotopic fidelity of ^{14}C measurements. The current AMS testing procedure for ^{14}C involves swiping the area of interest with a quartz filter which is dried and combusted in a sealed tube to produce CO_2 that then is reduced to graphite and measured. Problems that arise are cross contamination of equipment and that small samples may need to be diluted causing inconclusive results. We have developed a method using wet chemical oxidation and a gas ion source to help overcome these issues [1]. We use the method to test sites outside of our clean working areas for signs of elevated ^{14}C.

Fig. 1: *Swiping a door handle with quartz filter to check for elevated levels of ^{14}C.*

In 2014, more than 500 samples and standards from 6 institutions were analyzed. More than 100 above natural abundance (above modern) results were used to detect contamination point sources at 5 institutions. Point sources were connected to current and past use of ^{14}C enriched radiochemicals. Above modern samples had mean mass of 90 µg C with median fraction modern Fm = 2.2 and a maximum Fm > 3000. Below modern samples had a mean mass of 90 µg C with Fm of 0.8. Blanks had a mean mass of 20 µg C with Fm = 0.6 which was derived from the reagents and consumables.

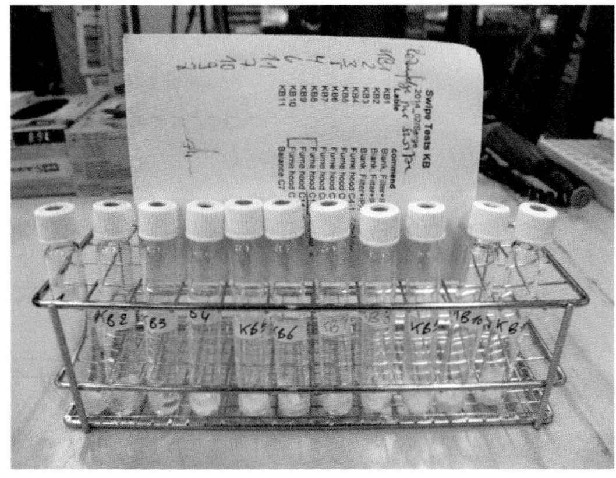

Fig. 2: *Contamination swipes prepared using the WCO method [1] ready for the MICADAS.*

We have developed a convenient and rapid method for testing for ^{14}C contamination on samples without dilution and minimizing cross contamination of equipment. We were able to identify problem areas and implement mitigation strategies. It was possible to clean areas with low levels of contamination to below modern levels. It was not readily possible to clean areas with ^{14}C values above Fm = 10. ^{14}C contamination was more widespread than originally thought. A direct injection method is being developed [2] to increase the throughput of samples analyzed on the MICADAS and minimize cross contamination of the gas interface.

[1] S. Lang et al., LIP Annual Report (2014), 60
[2] M. Seiler et al., LIP Annual Report (2014), 25

[1] *Geology, ETH Zurich*
[2] *Geochemistry and Petrology, ETH Zurich*

RAPID ^{14}C ANALYSIS OF DISSOLVED ORGANIC CARBON

Wet chemical oxidation for the preparation of DOC in non-saline waters

S. Lang[1,2], C. McIntyre, G. Früh-Green[1], S. Bernasconi[3], L. Wacker

The ^{14}C content of dissolved organic carbon (DOC) in rivers, lakes, and other non-saline waters can provide valuable information on carbon cycling dynamics in the environment. Currently, DOC is prepared for analysis with AMS using one of two offline methods. With the first method, samples are oxidized under vacuum using ultraviolet light. Using this method, samples can be analyzed at a rate of approximately one per day. With the second method, samples are freeze-dried in quartz tubes and combusted to CO_2 in the presence of cupric oxide, in a similar fashion to solid organic carbon samples.

We recently developed a method to determine the $\delta^{13}C$ content of DOC using wet chemical oxidation (WCO) [1]. The method has the benefit of low blanks and that samples can be prepared and analyzed rapidly, particularly when a headspace auto-sampler is employed. An automated headspace sampling interface was recently installed on the miniaturized radiocarbon dating system (MICADAS). We therefore tested the feasibility of extending the WCO procedure for the online radiocarbon analysis of DOC in freshwaters.

In brief, water samples were transferred into cleaned Exetainer® screw capped vials and acidified with phosphoric acid. Sodium persulfate was added as an oxidant and samples were purged with high-purity helium to eliminate any inorganic CO_2 from the vial. The samples were then heated to 100 °C for one hour to convert any sample DOC to CO_2.

For analysis, the samples were loaded into the carbonate handling system (CHS) of the MICADAS which was modified with a sparging needle. With this, helium could be bubbled through the water to displace all the sample CO_2 gas into the headspace. The headspace gas was then sent to the AMS for ^{14}C analysis.

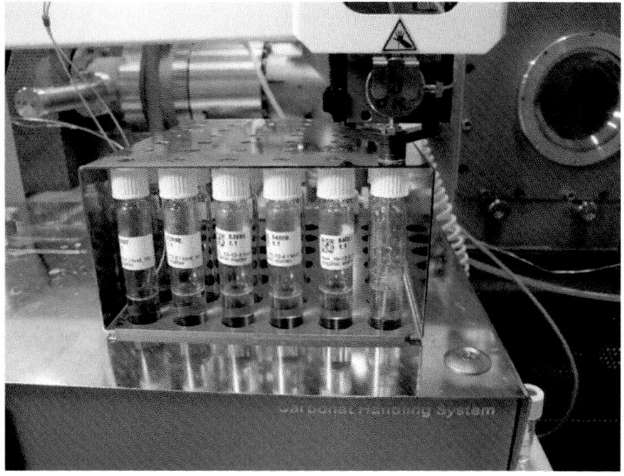

Fig. 1: *Online ^{14}C analysis of DOC samples using the WCO method.*

Five freshwater samples whose DOC $F^{14}C$ content had been assessed by other means were analyzed using the new WCO method. Three modern riverine samples with DOC concentrations between 200 – 550 µmol C/L had been analyzed using UV-oxidation at the National Ocean Sciences Accelerator Mass Spectrometry Facility (NOSAMS) in the United States. The $F^{14}C$ values determined by the new wet chemical oxidation method differed from the NOSAMS values by 0.007 to 0.065. Two other stream samples were analyzed by freeze-drying and quartz tube combustion and had $F^{14}C$ values of 0.168 and 1.071. Values determined by the WCO method differed by 0.038 and 0.005, respectively. In a recent application of the method, 22 samples plus 12 processing standards were prepared in one day, and analyzed online in a second day.

[1] S.Q. Lang et al., RCM, 25 (2011) 1

---

[1] *Geochemistry and Petrology, ETH Zurich*

[2] *Earth and Ocean Sciences, University of South Carolina, USA*

[3] *Geology, ETH Zurich*

COSMOGENIC NUCLIDES

COMBINING ^{10}BE AND *IN SITU* ^{14}C AT GRUEBEN GLACIER

Insights gained from cosmogenic nuclides on Holocene snow cover

C. Wirsig, S. Ivy-Ochs, N. Akçar[1], M. Lupker[2], K. Hippe, L. Wacker, C. Vockenhuber, C. Schlüchter[1]

The spatial and temporal extent of snow cover in a study area is often poorly known. Particularly in mountainous areas, however, shielding by snow can introduce significant uncertainties when reporting surface exposure ages derived from cosmogenic nuclides. Recently though, methods have been developed to constrain the extent of snow cover by combining cosmogenic ^{10}Be and *in situ* ^{14}C [1]. Essentially, the ^{10}Be/^{14}C ratio shifts during burial due to the different half-lives.

Fig. 1: *Picture of the bedrock ridge to the north of Lake Grueben taken on June 18th 2013. The three visible sample spots (red triangles) are clear while most of the ridge remains covered by snow. The cairn in the foreground stands roughly 50 cm tall.*

We collected four samples on a granite bedrock ridge at Grueben glacier in the central Swiss Alps that were analyzed for ^{10}Be, *in situ* ^{14}C and ^{36}Cl concentrations. The closest weather station located at Grimsel Hospiz (1970 m a.s.l.) reports average snow cover of roughly 150 cm during eight months per year for the last 66 years. Evidently, this might not be representative for the entire exposure period of roughly 10 ka. In addition, since our sample sites were specifically chosen on the top of exposed bedrock outcrops to minimize the influence of cover of any kind (Fig. 1), we expect less snow accumulation there than observed at the weather station.

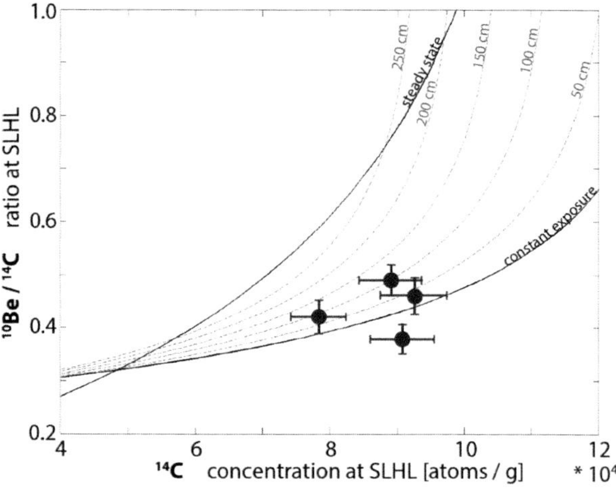

Fig. 2: *Multi-nuclide plot of ^{10}Be and in situ ^{14}C. Samples should plot between the lines of constant exposure and steady state. Scenarios with snow covering the samples for 8 months per year with an average thickness of 50 cm – 250 cm are represented by the dashed blue lines.*

Our analysis indicates a seasonal snow cover of less than 150 cm in agreement with the meteorological data (Fig. 2). The most likely values range from 50 cm – 100 cm during eight months per year. This result has significant implications for the calculated ages: an assumed average snow cover of 50 cm introduces a 9 % correction in exposure ages to the older.

[1] K. Hippe et al., Quat. Geochronol. 19 (2014) 14

---

[1] *Geology, University of Bern*
[2] *Geochemistry and Petrology, ETH Zürich*

HOLOCENE EROSION RECORDED BY AN ALPINE FAN

Decreasing post glacial erosion rates revealed through in-situ ^{10}Be

S. Savi[1], K.P. Norton[1,2], V. Picotti[3], N. Akçar[1], R. Delunel[1], F. Brardinoni[4], P. Kubik, F. Schlunegger[1]

Studies carried out in mountain environments have focused on exploring how the landscape's inheritance related to the erosional and depositional processes during the Last Glacial Maximum (LGM) has influenced sediment yields during the post-glacial periods. In this work [1] the stratigraphic record of a c. 3.5 km^2-large Holocene alluvial fan, located in the central-eastern Italian Alps, was used to reconstruct the erosional history of the adjacent source area in relation to a changing paleoclimate between the termination of the LGM and the modern situation. This was accomplished through the study of core material encountered on this fan that was analyzed for ^{14}C ages, the stratigraphy, and the ^{10}Be concentrations of the detrital material. In addition, a seismic section across the fan allowed the reconstruction of the sedimentary architecture and its evolution during the Holocene.

The results document a natural system where process rates have varied substantially during the Holocene, and where material has been supplied through fluvial and debris flow processes. From the LGM to the very early Holocene, a large sediment supply characterized by sediment accumulation rates on the fan up to 40 mm/yr is related to ice-melted discharge and climate variability, which conditioned high erosion rates in the catchment (Fig. 1). This flux then dramatically decreased toward interglacial values (^{10}Be-based denudation rates of 0.8 mm/yr at 5–4 cal kyr B.P.). Interestingly, the flux recorded in the fan material shows secondary peaks at 6.5 kyr and 2.5 kyr, when ^{10}Be-based erosion rates reached values of up to 13 mm/yr and 1.5 mm/yr, respectively. Paleo-denudation rates then decreased from ~33 mm/yr at the beginning of the Holocene to 0.42 mm/yr at 5 kyr, with peaks of ~6 mm/yr and 1.1 mm/yr at 6.5 kyr and 2.5 kyr.

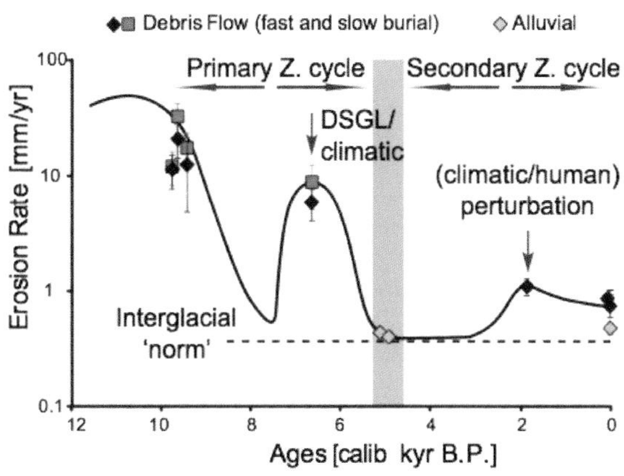

Fig. 1: *Pattern of erosion and sediment discharge, taken from [1].*

The pattern was interpreted in the sense that high-amplitude climate change was the most likely cause of the secondary peaks, but anthropogenic activities may also have contributed to this pattern [1]. The good correlation between paleo-sedimentation and paleo-denudation rates suggests that the majority of the deglaciated material destocked from the source area has been stored in the alluvial fan.

[1] S. Savi et al., GSA Bull. 126 (2014) 773

---

[1] *Geology, University of Bern*
[2] *Geography, Victoria University of Wellington*
[3] *Geology, University of Bologna*
[4] *Earth Sciences, University of Milano-Bicocca*

SEDIMENT TRANSFER IN A HIMALAYAN CATCHMENT

Insights from paired ^{10}Be and *in situ* ^{14}C measurements of river sands

M. Lupker[1], K. Hippe[1], L. Wacker, M. Christl, R. Wieler[1]

Cosmogenic nuclides in detrital river sediments have been widely applied to derive denudation rates and sediment fluxes across entire catchments [1]. Nuclides, such as ^{10}Be, allow the derivation of denudation rates integrated over several kyrs, but single isotopic systems often provide little information on the intricate dynamics that control the export of sediments from catchments. The quantification of sediment storage and recycling within catchments is nevertheless crucial for a better understanding of the variability of sediment fluxes and their implication for landscape evolution. The paired measurement of ^{10}Be along with cosmogenic, *in situ* ^{14}C in river sediments may provide new insights into sediment dynamics over kyr time scales for which other nuclides are not suitable [2,3].

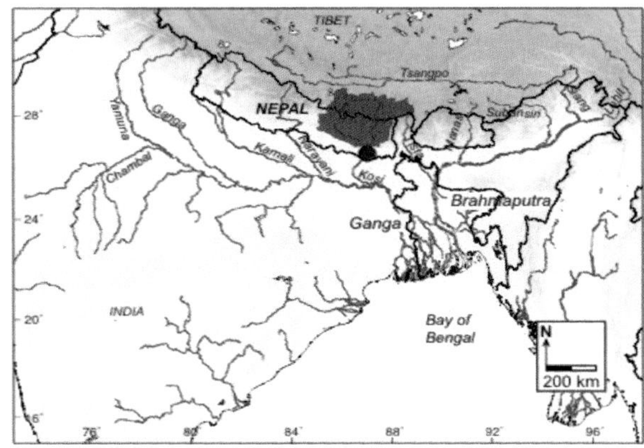

Fig. 1: *The Kosi catchment in eastern Nepal.*

In an effort to better understand the sediment dynamics in active orogens we combine *in situ* ^{14}C and ^{10}Be measurements from the Kosi basin (~53'000 km^2, Fig. 1). Our preliminary ^{14}C/^{10}Be data show apparent burial/storage ages of 14 to 21 kyr in the sediments currently exported by the river (Fig. 2).

These elevated burial ages suggest a larger storage component than previously thought in these catchments [4]. Possible biases associated with the use of ^{14}C/^{10}Be in sediments as a burial chronometer have to be considered. First, the short half-life of ^{14}C cannot be neglected and hence basin-wide denudation cannot be considered as a simple mixing of sediments from individually eroding surfaces, introducing bias towards higher apparent burial ages in most settings. Second, in steep environments, sediments supplied by deep-seated landslides carry a buried signature that should not be confounded with sediment storage in the catchment. The importance of both biases needs to be quantified carefully, before basin-wide storage can be quantified.

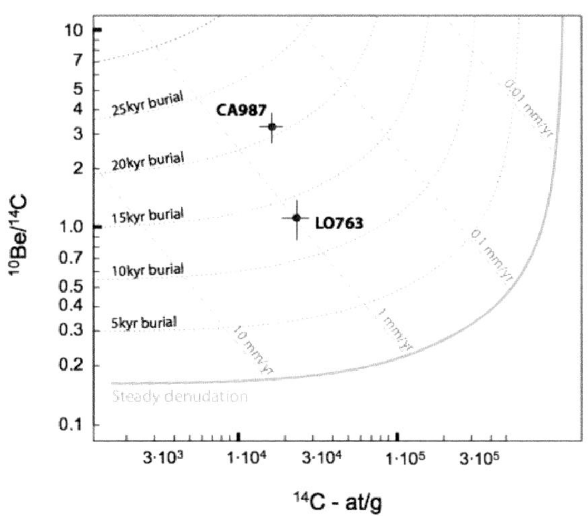

Fig. 2: *^{14}C–^{10}Be burial plot of two sediments exported by the Kosi River.*

[1] F. von Blanckenburg, EPSL 237 (2005) 462
[2] J.W. Lauer and J. Willenbring, JGR 115 (2010) F04018
[3] K. Hippe et al., Geomorph. 179 (2012) 58
[4] J.H. Blöthe and O. Korup, EPSL 382 (2013) 38

[1] *Geochemistry and Petrology, ETH Zurich*

THE ACTIVE ANDEAN THRUST FRONT (32°-32°30'S)

A combined structural, geomorphic and exposure dating approach

E. Garcia Morabito[1], R. Zech[2], C.M. Terrizzano[1], N. Haghipour[1], M. Christl, S.D. Willett[1], J.M. Cortes[3], V.A. Ramos[3]

For this study, we applied an innovative combination of structural and geomorphic investigations with surface exposure dating using cosmogenic nuclides (^{10}Be) in order to provide insights into the ongoing tectonic processes along the active deformation front of central western Argentina, one of the most active zones of thrust tectonics in South America [1].

Fig. 1: *Active Andean thrust front in Central Western Argentina and sample locations.*

We concentrated on active reverse faults and alluvial terraces at two key locations (Fig. 1), where rates of uplift and shortening are still unknown due to the lack of ages for the abandoned and uplifted Quaternary deposits. Absolute ages of the preserved aggradation levels might also provide valuable information about past environmental conditions. We used several geomorphic approaches, including drainage analysis, DGPS surveys, and surface reconstruction, and combined them with ^{10}Be surface exposure dating.

Three deformed levels are preserved in the Cerro Salinas area (Fig. 2). Here, we obtained preliminary surface exposure ages for amalgamated pebbles, and via depth profiles on sand samples. These new measurements yield terrace abandonment ages between 21 and 90 Ka.

Age distribution allows determining a chronology of uplifted surfaces and quantifying rates of deformation and incision. Preliminary data suggest recent vertical uplift or incision rates of 0.3-0.5 mm/yr. On-going work aims to obtain additional ^{10}Be exposure ages and long-term deformation rates.

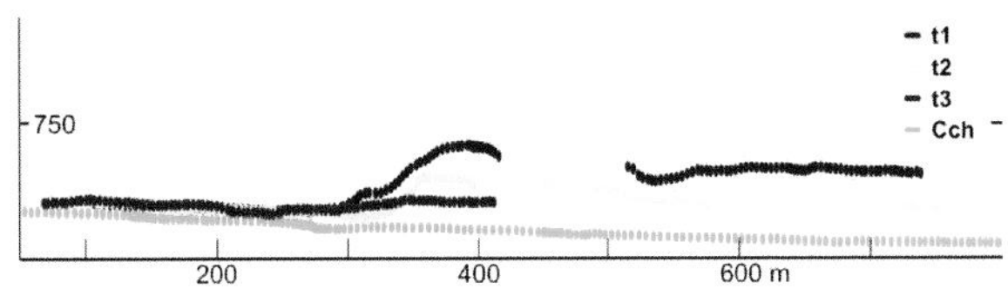

Fig. 2: *DGPS topographic profiles surveyed on the northern terraces of the Cerro Salinas anticline projected over a line perpendicular to the fold axis.*

[1] J. Vergés et al., J. Geo. Res. 112 (2007)

[1] *Geology, ETH Zurich*
[2] *Geography, University of Bern*
[3] *IGEVA, University of Buenos Aires*
[4] *IDEAN, University of Buenos Aires*

NEOTECTONICS OF THE CENTRAL ANDES

Rates of Quaternary uplift based on ^{10}Be surface exposure dating

C.M. Terrizzano[1], R. Zech[2], E. Garcia Morabito[1], M. Yamin[3], N. Haghipour[1], L. Wüthrich[2], M. Christl, S. Willett[1], J.M. Cortes[4]

The active Andean deformation front in the Argentinian forelands between 28° and 33°S is currently concentrated at the boundaries of a main morphotectonic unit: The Precordillera. Most studies so far have focused on estimating slip and uplift rates along its eastern margin [1, 2]. Rates of uplift and shortening on the reverse faults that uplift its western flank remain largely unknown.

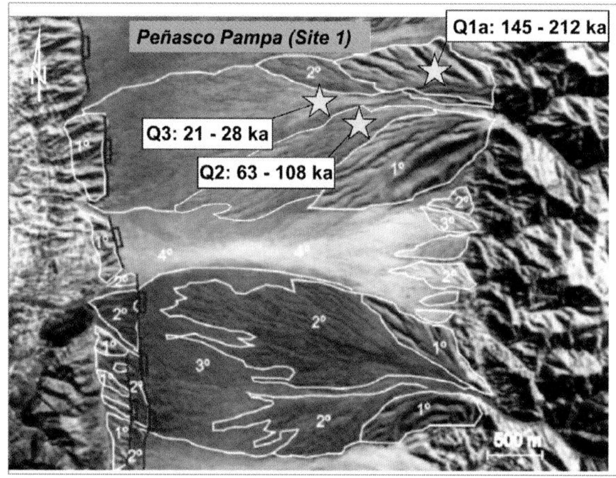

Fig. 1: *Alluvial fans in the Western Precordillera affected by Quaternary faults.*

We combined structural and geomorphic investigations with ^{10}Be surface exposure dating in order to establish a numerical chronology for four deformed alluvial fan surfaces and to estimate uplift rates of the Barreal block (31°30′-31°53′S/69°20′W) in the western Precordillera (Fig. 1). Surface exposure ages were determined for a few large boulders, amalgamated pebbles, and via depth profiles on sand samples. Boulder ages range from 145 to 212 ka for the oldest well-preserved fan remnants (Q1a, n=3), from 63 to 108 ka (Q2, n=3) and 21-28 ka (Q3, n=2, amalgamated pebbles yield ages range from 106 to 127 ka for the oldest fan surface (Q1b, n=79). The depth profiles yield minimum ages (assuming negligible erosion) of 120 ka (Q1a as well as Q1b) (Fig. 2) and 79 ka (Q2) and are thus mostly in good agreement with boulder ages.

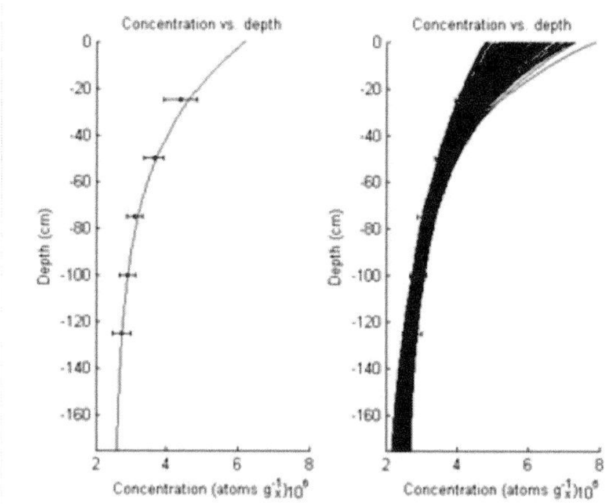

Fig. 2: *Concentration vs depth plot of Q1a.*

Uplift rates of 0.32 mm/yr and 0.83 mm/yr can be estimated for the eastern and western front of the Barreal block respectively. This provides the first numerical constraints for the neotectonic activity of this region. The exposure ages may tentatively be interpreted to document fan formation at times of globally low temperatures and glacial maxima (Marine Isotope stages 2, 4 and 6, respectively).

[1] S. Schmidt et al., Tectonics 30 TC5011 (2011)
[2] L. Siame et al., GJI 150 (2002) 241

---

1 Geology, ETH Zurich
2 Geography, University of Bern
3 SEGEMAR, Argentina
4 IGEBA, University of Buenos Aires

EROSION RATES IN SOUTHERN TAIWAN

Basin-wide erosion rates estimated from ^{10}Be in modern river sands

C.Y. Chen[1], S.D. Willett[1], N. Haghipour[1], M. Lupker[2], M. Christl

Taiwan is a young and rapidly evolving orogen built by an oblique arc-continent collision. The high tectonic convergence rates coupled with sub-tropical climate and frequent typhoons result in extremely high uplift and exhumation rates.

In order to understand the evolution of this mountain building, many efforts have been made to characterize the uplift or erosion rates of Taiwan on different timescales except the millennial timescale.

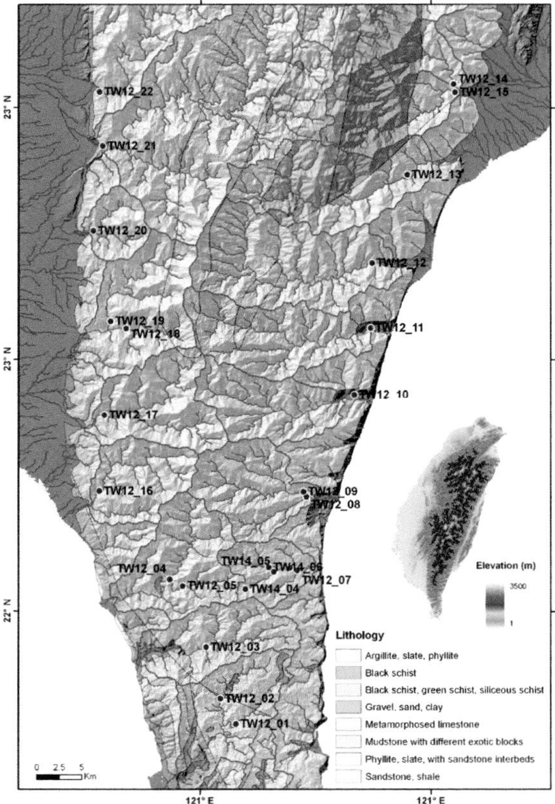

***Fig. 1:** Geological map and distribution of samples collected in Southern Taiwan. Insert shows the colored shaded-relief map of Taiwan.*

The extremely high erosion rates and the concomitant very low nuclide concentrations make Taiwan a difficult area to measure erosion rates with ^{10}Be concentrations in river sediments. Recently, ^{10}Be-based erosion rates for several large catchments in western, eastern and northern Taiwan have been published [1][2].

Millennial timescale erosion rates in Southern Taiwan have not been determined yet. For this study, we collected modern river sands from each main catchment in Southern Taiwan (Fig. 1) in order to characterize the erosion rates and understand how the landscape of Taiwan evolved spatially through time.

***Fig. 2:** Photographs of some of sample sites.*

[1] F. Derrieux et al. J. Asian Earth Sci. 88 (2014) 230

[2] L.L. Siame et al. Quat. Geochron. 6 (2011) 246

[1] *Geology, ETH Zurich*

[2] *Geochemistry and Petrology, ETH Zurich*

^{10}BE DEPTH PROFILE DATING IN THE SWISS MIDLANDS

Deposition ages versus erosion

L. Wüthrich [1] R. Zech[1,2], N. Haghipour[2], C. Terrizzano[1,2], S. Ivy-Ochs, M. Christl, C. Gnägi[3], H. Veit[1]

During the Pleistocene, glaciers advanced repeatedly from the Alps into the Swiss Midlands. The exact extents and timing are still under debate. Decalcification depths, for example, increase from west to east in the western Swiss Midlands and have been interpreted to indicate that the Rhone glacier may have been less extensive during the global Last Glacial Maximum (LGM) at 20 ka than has been assumed so far [1].

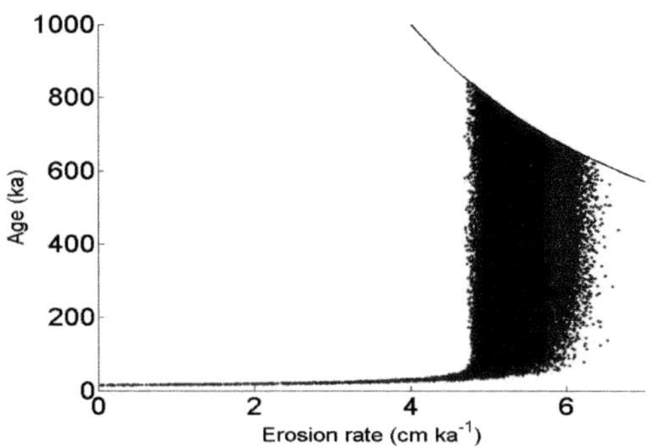

Fig. 1: *Age vs. erosion plot for the Niederbuchsiten Site. Red dots mark the possible solutions, blue dots the 100 best χ^2 fits. The black curve marks the total erosion threshold of 40 m.*

We applied ^{10}Be depth profile dating [2] on till at five locations in the western part of Switzerland. Two of them lie outside of the assumed LGM extent of the Rhone glacier (Niederbuchsiten, St. Urban), two inside the extent of the LGM Rhone glacier (Steinhof, Deisswil) and one profile was taken from the Bern stadial (LGM) of the Aare glacier [3]. All surface concentrations are relatively low and indicate considerable erosion. Without constrains for age and erosion, depth profile dating yields ages between roughly 15 ka up to more than 1 Ma for the profiles in St. Urban, Niederbuchsiten and Deisswil; whereas the profiles in Steinhof and Bern yield only last glacial ages.

The wide range of possible ages illustrates that independent estimates for erosion are needed to precisely determine the deposition ages of the investigated tills. However, at this point, we interpret the best model fits to our depth profile concentrations as tentative verification of the assumed LGM extent [3,4]. The spatial patterns of decalcification depths and soil development in the Swiss Midlands merit further evaluation.

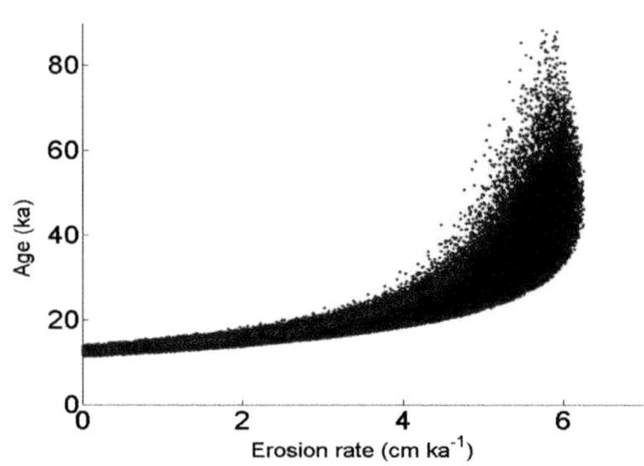

Fig. 2: *Age vs. erosion plot for the Steinhof site. Red dots mark the possible solutions, blue dots the 100 best χ^2 fits.*

[1] T. Bitterli et al., Geologischer Atlas der Schweiz, Blatt 1108 (2011)

[2] A.J. Hidy et al., G^3 (2010) 11

[3] A. Bini et al., Switzerland during the Last Glacial Maximum (1:50000) Swisstopo (2009)

[4] S. Ivy-Ochs et al., Eclogae Geologicae Helvetiae (2004) 47

[1] *Geography, University of Bern*

[2] *Geology, ETH Zurich*

[3] *Weg>Punkt, Herzogenbuchsee*

INITIATION TIME OF THE DRIEST INSTABILITY AT ALETSCH

Surface exposure dating with ^{10}Be on a landslide head-scarp

L. Grämiger[1], J.R. Moore[2], S. Ivy-Ochs, M. Christl, S. Loew[1]

The temporal distribution of landslides following deglaciation is one of the most important components of the paraglacial concept [1]. To establish better understanding of links between deglaciation and initiation of landslide processes, it is necessary to constrain initial failure ages of paraglacial slope instabilities. Here we present a case study of a landslide from an adjacent rock slope at the Grosser Aletsch glacier, Switzerland.

The Driest instability is located on the west slope of the Grosser Aletsch glacier tongue within the Aar granitic massif (Fig. 1). The steep head-scarp of the landslide lies around 50 m below an Egesen stadial moraine [2,3]. The Little Ice Age moraine is clearly visible in the disintegrated and unstable landslide body, around 100m below the head-scarp. The landslide shows recent movement visible by freshly exposed rock and evaluated using remote monitoring [4].

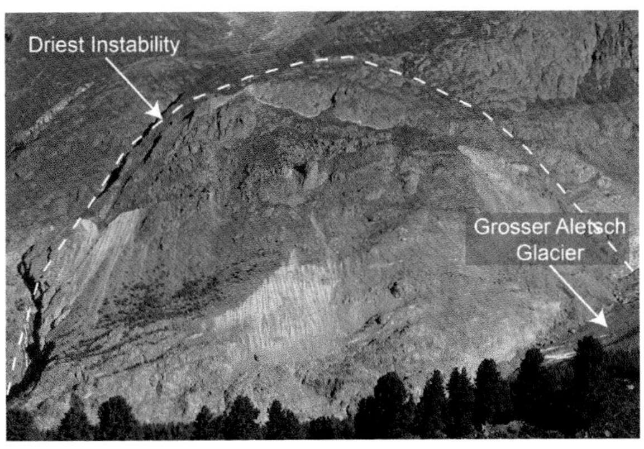

Fig. 1: *Driest instability above the Grosser Aletsch glacier.*

To constrain the initial age of the Driest landslide, five bedrock samples aligned along the scarp were taken for cosmogenic nuclide surface-exposure dating with ^{10}Be (Fig. 2). The top sample was taken from glacially polished bedrock, whereas the other samples are from the scarp. The calculated exposure dates show an early Holocene initial failure age, and indicates that initial deformation along the head-scarp followed Egesen deglaciation.

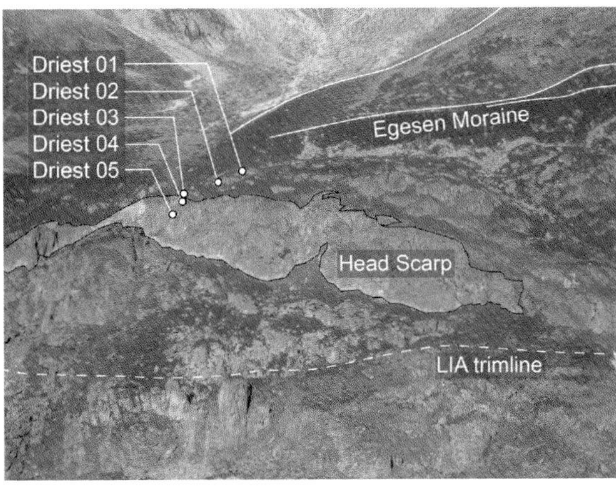

Fig. 2: *Sample locations at the Driest instability with Little Ice Age and Egesen glacier extents.*

[1] C.K. Ballantyne, Quat. Sci. Rev. 21 (2002) 1935
[2] M.A. Kelly et al., J. Quat. Sci. 19 (2004) 431
[3] I. Schindelwig et al., J. Quat. Sci. 27 (2012) 11.
[4] A. Kääb, ISPRS J. of Photogrammetry & Remote Sensing 57 (2002) 39

---

1 Geology, ETH Zurich
2 Geology and Geophysics, University of Utah, USA

AGE OF THE MONTE PERON ROCK AVALANCHE

^{36}Cl dating of landslide deposits in the Belluna Valley, Italy

S. Ivy-Ochs, S. Martin[1], C. Vockenhuber, V. Alfimov, N. Surian[1], M. De Zorzi[1], G. Carugati[2], P. Campedel[3], M. Rigo[1], A. Viganò[4]

In the Late Pleistocene, in the southeastern side of the Alps (Venetian region, Italy), after the Piave glacier tongues had retreated from the end moraine system areas back towards the Dolomitic region, large landslides took place in the Belluno Prealps.

Fig. 1: *The Monte Peron rock avalanche deposits (foreground) and the release area (background).*

The aim of this study is to determine when the Monte Peron landslide occurred using ^{36}Cl surface exposure dating. In the Monte Peron area, debris reached the bottom of the Cordevole valley forming several tens of meters thick deposits with a volume of about 100 x 10^6 m^3. The landslide covers an area of roughly 20 km^2, extending several kilometers from the main scarp (Fig. 1).

According to previous work, the main event occurred during the first phases of deglaciation, between 17,000 and 15,000 years BP (Pellegrini et al., 2006). Our ^{36}Cl exposure dating gave historical ages for several boulders on the lansdslide, both near the Monte Peron scarp and far from the main scarp (Fig. 2). According to these exposure ages we cannot exclude the hypothesis that earthquakes related to the Belluno fault could have played a key role for triggering of the landslide. The main gravitational event took place in historical times rather than during the deglaciation.

Fig. 2: *Extent of Monte Peron rock avalanche. Locations of boulders sampled for ^{36}Cl dating are indicated. Width of area shown ca. 5 km.*

[1] G.B. Pellegrini, N. Surian, D. Albanese, Geogr. Fis. Dinam. Quat. 29 (2006) 185

[1] *Geoscience, University of Padua, Italy*
[2] *National Institute of Oceanography and Experimental Geophysics, CRS, Udine, Italy*
[3] *Geological Survey of the Province of Trento, Italy*
[4] *Chemical and Environmental Sciences, University of Insubria, Como, Italy*

CLIMATE CHANGE RECORDED BY FLUVIAL TERRACES

Climate-driven sediment discharge revealed through in-situ ^{10}Be

T. Bekaddour[1], F. Schlunegger[1], H. Vogel[1], R. Delunel[1], K.P. Norton[1,2], N. Akçar[1], P. Kubik

It is a relatively complicated task to understand landscape response, especially erosional response, to climate change. In this context, the formation of terrace sequences has been related to climate-driven disturbances in sediment flux where landsliding resulted in the temporary storage of sediment in the channel network, thereby constructing a terrace level. In this study [1] we analyzed such a situation offered by the Quaternary cut-and-fill terrace sequences in the Western Andean Escarpment of Peru between 5°S and 20°S latitude.

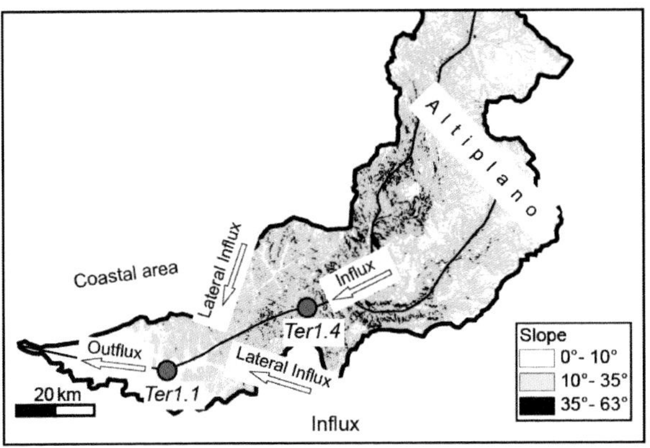

Fig. 1: *Pisco drainage basin, and illustration of concept for the calculation of the ^{10}Be-based sediment budget. Ter1.1 and Ter1.4 represent sites where the Minchin terrace deposits were sampled for in-situ ^{10}Be in river sediment [1].*

This analysis focused on the Pisco valley at lat. 13°S that covers an area of 4300 km^2 and extends ~200 km in an east-west direction (Fig. 1). It hosts three terrace sequences that have been dated to 36-48 ka (Minchin pluvial period) and 15-26 ka (Tauca pluvial period) and younger ages through optically stimulated luminescence (OSL) techniques [2]. Building on the high-resolution chronological dataset [2] a ^{10}Be-based sediment budget was established (Fig. 1) for the cut-and-fill terrace sequences in this valley, with the aim to learn more about orbitally forced changes in precipitation and erosion.

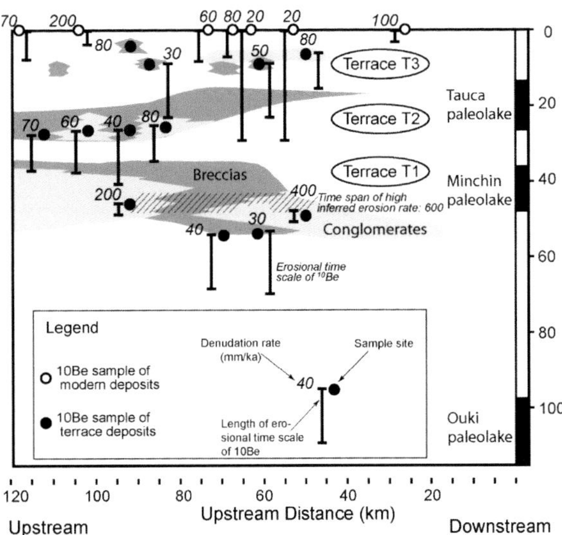

Fig. 2: *Results showing the erosion rate pattern recorded by ^{10}Be concentrations in the three terrace deposits and in the modern channels [1].*

The results (Fig. 2) show that the Minchin period was characterized by an erosional pulse along the Pacific coast (downstream reach) where denudation rates reached values as high as 600±80 mm/ka for a relatively short time span lasting a few thousands of years. This contrasts to the younger pluvial periods and the modern situation when ^{10}Be-based sediment budgets register nearly zero erosion at the coast.

[1] T. Bekaddour et al., Earth Planet. Sci. Lett. 390 (2014) 103

[2] D. Steffen et al., Geology 37 (2009) 491

[1] *Geology, University of Bern*

[2] *Geography, Victoria University of Wellington*

RAPID THINNING OF AN EAST ANTARCTIC GLACIER

Exposure dating and modelling the deglaciation of Mackay Glacier

R.S. Jones[1], K.P. Norton[1], A.N. Mackintosh[1], N.R. Golledge[1,2], C.J. Fogwill[3], P.K. Kubik and M. Christl

We investigate past thinning and dynamics of Mackay Glacier, an outlet of the East Antarctic Ice Sheet, using surface exposure dating and flow-line modelling. Our ^{10}Be-drived chronology of erratic cobbles, collected in altitudinal transects extending above the modern ice surface, records surface lowering from the Last Glacial Maximum to present-day (Fig. 1).

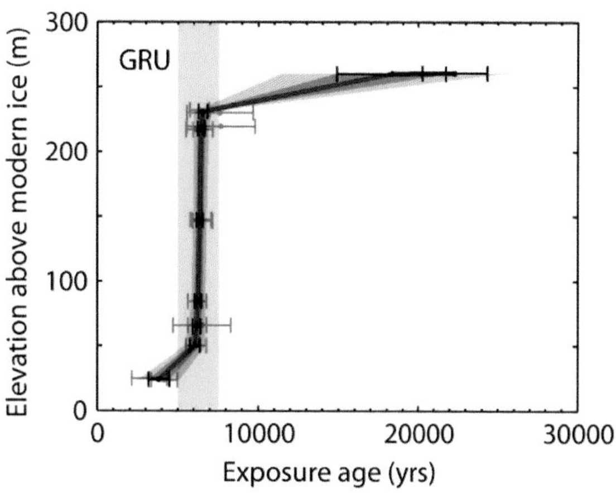

Fig. 1: *^{10}Be-derived thinning histories from Gondola Ridge. Raw ages (black) have been age-modelled using a Bayesian framework (blue). Transects show thinning from ~6.8 to 6 ka, with rates determined from Monte Carlo regression analysis (shown with mean and 2σ).*

Most notably, four transects reveal an episode of rapid thinning during the early-mid Holocene, at a time of relative climatic and oceanic stability. Statistical analysis reveals that >180 m of thinning occurred at rates similar to Pine Island, Thwaites and Totten Glaciers today, and persisted for at least ~400 years.

In an attempt to understand the mechanism of recorded rapid thinning, we simulated the glacier using a glacier flow-line model (Fig. 2). We find that modelled retreat of the grounding-line is initially gradual but then accelerates through an overdeepened trough on the inner continental shelf, irrespective of the forcing applied. Simulated thinning at our transects also dramatically accelerates during this time, replicating the different rates and magnitudes of surface lowering recorded in our chronology.

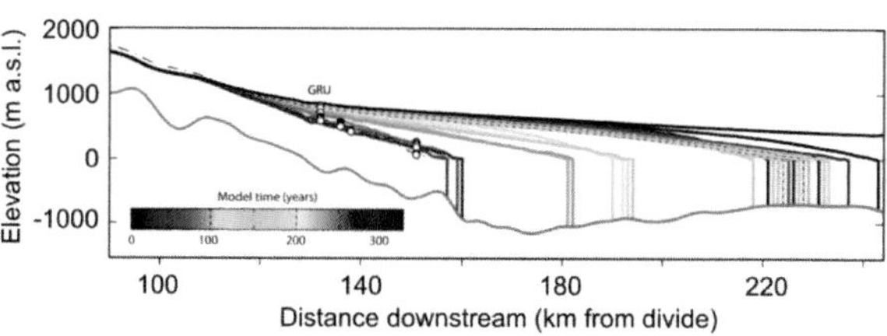

Fig. 2: *An evolving surface profile is shown at 10-year intervals for one model run (-1.75 °C ocean temperature forcing scenario).*

Using a combination of surface exposure dating, statistical analysis and glacier flow-line modelling we conclude that: (1) rapid Holocene thinning at Mackay Glacier, comparable to modern Antarctic thinning rates, persisted for >400 years and resulted in 100s of metres of ice loss; (2) this episode occurred 100-1000s of years after climatic changes ceased; (3) recorded thinning likely resulted from marine ice sheet instability as the glacier retreated into an overdeepened trough on the inner continental shelf.

---

[1] *Victoria University of Wellington, New Zealand*
[2] *GNS Science, New Zealand*
[3] *University of New South Wales, Australia*

COSMOGENIC NUCLIDE-DERIVED DENUDATION RATES

Rates over space and time from European rivers and terrace sequences

M. Schaller[1], T.A. Ehlers[1], P.W. Kubik, M. Christl, C. Vockenhuber, T. Stor[2], J. Torrent[3], L. Lobato[4]

The Earth surface is subjected continuously to denudation due to changes in tectonics, climate, and biotic activity. Determining denudation rates over space and time is not always easy. Cosmogenic nuclide concentrations in active river sediment and river terrace deposits contain information about catchment-wide denudation rates and paleo-denudation rates, respectively.

Temporal and spatial variations in denudation across Europe as a function of climate change are investigated. Paleo-denudation rates are determined from in situ-produced cosmogenic ^{10}Be and ^{26}Al measured in river terraces from four river catchments with different latitudes across Europe.

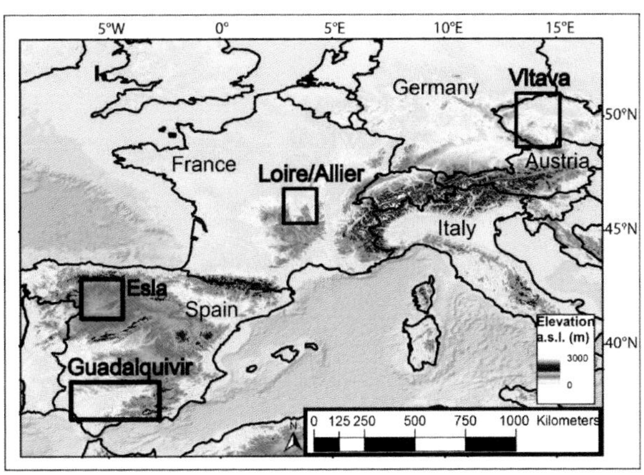

Fig. 1: Four European locations with terrace sequences analyzed for cosmogenic nuclide-derived denudation rates and paleo-denudation rates.

Cosmogenic nuclide-derived denudation rates from active river sediments are being determined from catchments in southern and northern Spain (Guadalquivir and Esla, respectively), central France (Allier and Loire), and the Czech Republic (Vltava).

The cosmogenic nuclide-derived denudation rates of the Guadalquivir range from 34 to 42 mm/ka. In the upper and lower course of the Esla river system, the respective denudation rates are 50 mm/ka and 30 mm/ka. For the Allier, denudation rates recalculated from measurements by [1] are around 40 mm/ka. The denudation rates of the Vltava and the Elbe are around 30 mm/ka with the Elbe at 38 mm/ka.

Fig. 1: Sampling of terrace material from the Vltava river (Czech Republic) for isochron burial dating.

In order to determine paleo-denudation rates, the terrace deposition ages need to be known. The method of isochron burial dating [2] is currently applied on several terraces in the four river catchments.

[1] Schaller et al., EPSL, 188 (2001) 441

[2] G. Balco and C.W. Rovey II, American J. of Sci., 308 (2008) 1083

[1] *Geosciences, University of Tübingen, Germany*

[2] *Geology, Charles University, Czech Republic*

[3] *Agronomy, University of Córdoba, Spain*

[4] *Geology, University of León, Spain*

PEDOGENESIS OF BELGIAN LOESS-DERIVED SOILS

Meteoric ^{10}Be inventories derived from authigenic ^{10}Be/^{9}Be ratios

V. Vanacker[1], B. Campforts[2], G. Govers[2], E. Smolders[2], S. Baeken[2], M; Christl

Soils developed on loess deposits are widespread in Europe, and provide fertile soils for agriculture and forestry. In the Belgian loess belt, the youngest loess accumulation period lasted most likely until the Heinrich 2 event (25 - 20 ka). Soil genesis is mainly characterized by decarbonisation followed by clay migration [1].

Fig. 1: *Soil profile (90 masl) developed in loess deposits in the Bertem Forest, located about 20 km east of Brussels.*

In this study, we use meteoric ^{10}Be as a tracer of soil development. One undisturbed soil profile was analysed in the Bertem forest, at the northern edge of the Belgian loess belt. The soil profile was dug down to the calcareous loess of the C horizon (610 cm). Dry soil bulk densities ranged from 0.60 to 1.84 g cm^{-3}, with highest densities measured for the Bt argic horizon.

Eight soil samples (taken at depths of 7, 14, 27, 50, 72, 73, 115 and 550 cm) were analyzed for their authigenic ^{10}Be/^{9}Be ratio and ^{9}Be concentration. Two fully processed blanks were also measured to assess potential contamination with ^{10}Be or ^{9}Be during the preparation of the carrier-free samples.

Carrier-free ^{10}Be/^{9}Be ratios were measured using the TANDY AMS facility (600 kV) at ETH Zurich [2], and ^{9}Be concentrations were measured independently using an Agilent 7700x ICP-MS at KULeuven (Belgium). The mean deviation of the ^{9}Be concentration from certified samples was 4%, while the total error on the natural ^{10}Be/^{9}Be after background corrections ranged between 2.4 and 5.2%.

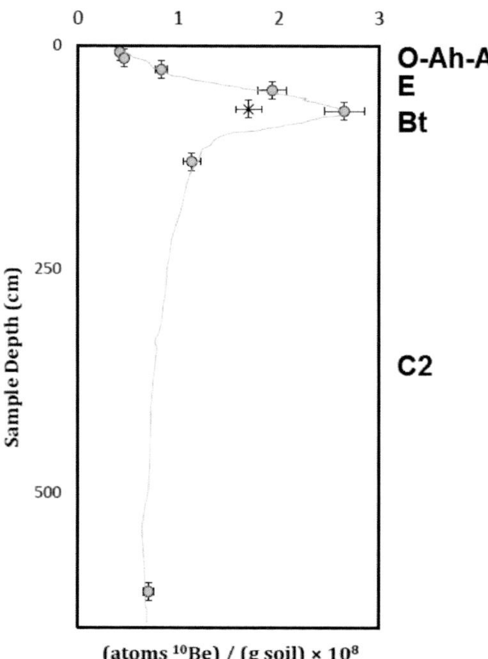

Fig. 2: *Plot of meteoric ^{10}Be concentration as a function of depth in the undisturbed soil profile.*

Our results show a systematic variation of ^{10}Be with depth. Concentrations of ^{10}Be range from 4.14×10^{7} to 2.65×10^{8} atoms ^{10}Be (g soil)$^{-1}$. The concentrations are systematically higher in the clay-enriched Bt and Btx horizons, and lower in the organo-mineral horizons.

[1] V. Brahy et al., Eur. J. Soil Sci. 51 (2000) 1
[2] M. Christl et al., Nucl. Instr. & Meth. B 294 (2013) 29

[1] *Earth & Life Institute, University of Louvain, Belgium*
[2] *Soil and Water Management, KULeuven, Belgium*

COSMOGENIC NUCLIDES IN LARGE LOWLAND BASINS

In situ vs. meteoric ^{10}Be: Different behavior but same outcome

H. Wittmann[1], F. von Blanckenburg[1], N. Dannhaus[1] J. Bouchez[1,2], J.L. Guyot[3], L. Maurice[4], J. Gaillardet[2], M. Christl

Meteoric ^{10}Be concentrations increase from the Andes to the lowland area of the Amazon as opposed to *in situ*-^{10}Be concentrations that are uniform over the same distance [1,2] (Fig. 1). However, results from our new erosion and weathering proxy, the meteoric ^{10}Be to stable ^{9}Be ratio that can be measured in reactive phases extracted from bedload or suspended river sediment (yielding $[Be]_{reac}$), indicate that denudation rates from both nuclides are very similar (Fig. 2) except for some disparity in the Andes, attributed to too high meteoric fluxes.

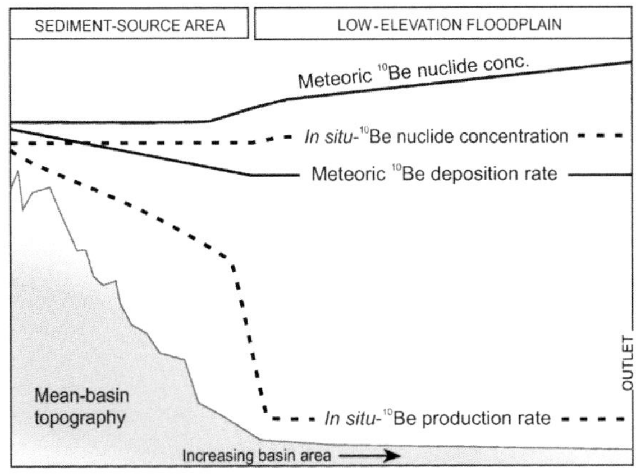

Fig. 1: *Concept of the behavior of in situ vs. meteoric ^{10}Be from high-relief to low-relief area.*

Using suspended sediment samples integrated over depth profiles, considered to be more representative than bedload, an increase in $[^{10}Be]_{reac}$ from 18×10^{6} to 32×10^{6} atoms/g is recorded over this distance, whereas $[^{9}Be]_{reac}$ stays roughly constant. In total, an increase in the $(^{10}Be/^{9}Be)_{reac}$ by a factor of approx. three is recorded, too. At the same time, the meteoric depositional flux decreases by ca. 30% across the basin [3]. Resulting $(^{10}Be/^{9}Be)_{reac}$-derived denudation rates only decrease slightly, similar to those from *in situ* ^{10}Be, across the basin, although $[^{10}Be]_{reac}$ and $(^{10}Be/^{9}Be)_{reac}$ increase.

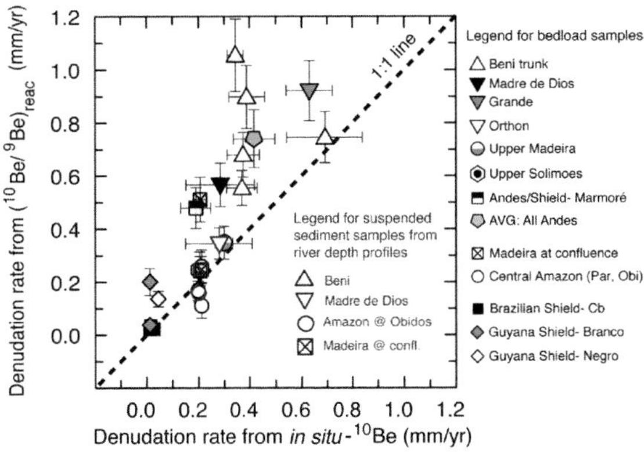

Fig. 2: *$(^{10}Be/^{9}Be)_{reac}$-derived versus in situ-derived denudation rates in the Amazon basin.*

In contrast, *in situ* ^{10}Be concentrations show only a minor increase of 5.2×10^{4} to 6.4×10^{4} atoms/g from the Andes to the central lowlands [2]. During transfer and storage in lowlands, decay or additional production of ^{10}Be may ensue, depending on storage depth and its duration [1]. However, at the surface of low-lying floodplains, *in situ* ^{10}Be only slowly accumulates due to low production rates (Fig. 1). Production decreases for *in situ* ^{10}Be by a factor of four across the lowlands [2]. As a result, the *in situ* ^{10}Be concentration does not increase across the lowlands.

Regarding behavior in lowlands, the major difference between *in situ* and meteoric ^{10}Be thus lies in their respective atmospheric scaling that then leads to similar denudation rates.

[1] Wittmann et al., Geomorph., 109 (2009) 246
[2] Wittmann et al., GSA Bull., 123 (2011) 934
[3] Heikkilä et al., Space Sci Rev., 176 (2013) 321

[1] *Helmholtz Centre and GFZ, Potsdam, Germany*
[2] *IPG Paris, Université Paris Diderot, CNRS, France*
[3] *IRD, Lima, Peru*
[4] *UPS (SVT-OMP), LMTG, Toulouse, France*

ISOTOPIC EVOLUTION OF THE DEEP ARCTIC OCEAN

Radiogenic isotopes reflect climatically driven weathering inputs

V.Dausmann[1], M. Frank[1], C. Siebert[1], M. Christl, J.R. Hein[2]

The isotope evolution of deep waters in the Arctic Ocean (2200-3600 m) was reconstructed from three hydrogenetic ferromanganese crusts dated using cosmogenic ^{10}Be (i.e. ^{10}Be/^{9}Be).

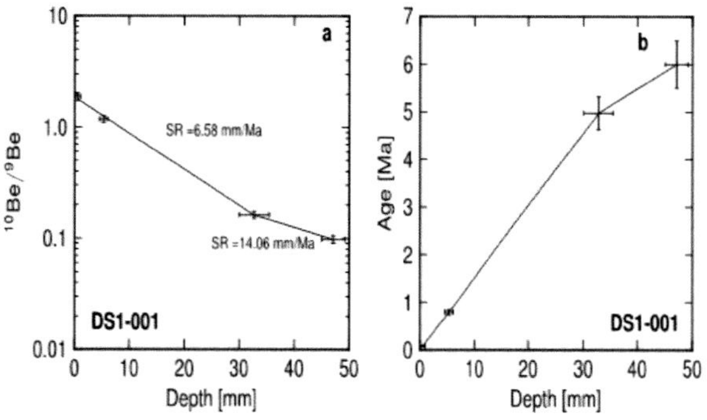

Fig. 1: *Example for ^{10}Be/^{9}Be profile (a) and age-depth relation (b) in crust DS1-001.*

The ^{10}Be/^{9}Be signature of the crust surfaces integrates over the past several 100'000 years and is lower (~2×10^{-8}) than expected from deep-water measurements (~7×10^{-8}) but consistent with previously obtained data from Arctic marine sediments [1]. This supports undisturbed present-day growth surfaces as well as a lower deep-water ^{10}Be/^{9}Be signal for most of the Late Quaternary. The Nd isotope time series of the three crusts show a pronounced trend to less radiogenic values from $\varepsilon_{Nd(Hf)}$ of -8.5 (+4) starting at ~4 Ma to -11.5 (-3) at the surfaces (Fig. 2), agreeing well with present day deep-water values.

The main finding of this study is that climatically controlled changes in weathering intensity and regime on the North American continent have controlled the isotopic evolution of Arctic Deep Water (ADW). From these records it is deduced that enhanced weathering inputs linked to the onset of glacial conditions reached ADW as early as 4 million years ago.

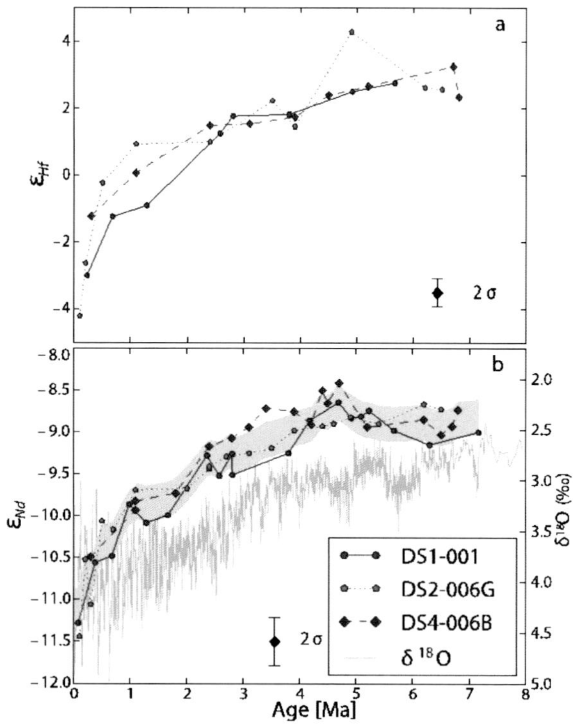

Fig. 2: *Radiogenic Hf (a) and Nd (b) (average in grey) isotope evolution of the deep Arctic Ocean over the last 8 Myr. Blue line in (a) represents the evolution of the global ice volume.*

These findings have implications for the reconstruction of the climatic evolution of the northern high latitudes. Furthermore comparison of our new data to previously obtained results for Arctic Intermediate Water on the Lomonosov Ridge demonstrate that, unlike today, pronounced isotopic differences between the water masses from different depths and areas of the Arctic Ocean prevailed for most of the past 7 million years.

[1] E. Sellén et al., Global Planet Change 68 (2009) 38

---

[1] *GEOMAR Helmholtz Centre for Ocean Research, Kiel*
[2] *United States Geological Survey (USGS)*

ANTHROPOGENIC RADIONUCLIDES

THE OCTOBER 2014 CRUISE ALONG THE COAST OF JAPAN

Tracking radioactive releases 2 years after the Fukushima accident

N. Casacuberta, M. Christl, P. Masqué[1], K.O. Buesseler[2], J. Vives i Batlle[3], J. Nishikawa[4], M. Aoyama[5]

In October 2014, a cruise along the coast of Japan took place on-board the R/V Shinsei Maru (Fig. 1). It started on 17th October in Aomori (North of Japan) and ended the 27th October in Yokosuka Port (South of Tokyo). The cruise track consisted in 15 stations located off the coast of the Fukushima Prefecture.

This cruise is organized within the remit of the EU FRAME project (The impact of recent releases from the Fukushima nucleaR Accident on the Marine Environment), underpinned by a consortium of six international partners: Universitat Autònoma de Barcelona, Woods Hole Oceanographic Institution, Belgian Nuclear Research Center, ETH Zurich, Tokai University and Fukushima University.

Fig. 1: *Research Vessel Shinsei Maru at Yokosuka Port (South of Tokyo).*

The aim of FRAME is to understand the sources, fate, transport, bioaccumulation and associated impact of radionuclides from the Fukushima Dai-ichi Nuclear Power plants (NPP). For this purpose, artificial radionuclides (^{137}Cs, ^{134}Cs, ^{90}Sr, Pu-isotopes, ^{236}U and ^{129}I) are being analysed in seawater, sediments, fish, macroalgae and plankton.

Our group (LIP) was directly involved in the sampling campaign. A total of 40 seawater and sediment samples were collected, for the analysis of ^{236}U and Pu-isotopes (Fig 2.). Radiocaesium and ^{90}Sr isotopes will be analysed in a similar number of samples at Woods Hole Oceanographic Institution and the Universitat Autònoma de Barcelona, respectively.

Fig. 2: *Sampling seawater off the Coast of Japan. The Fukushima NPP is visible in the background.*

All data will be gathered together and used in models. Reverse transfer modeling work performed by the Belgian Nuclear Research Center, will allow for improved transfer and kinetic parameters between the biota and the surrounding environment.

---

[1] *Universitat Autònoma de Barcelona*
[2] *Woods Hole Oceanographic Institution*
[3] *Belgian Nuclear Research Center*
[4] *Tokai University*
[5] *Fukushima University*

NEW DATA OF ^{236}U AND ^{129}I IN THE ARCTIC OCEAN

Constraining the sources of ^{236}U and ^{129}I in the Arctic Ocean

N. Casacuberta, M. Christl, C. Vockenhuber, C. Walther[1], M. R. Van-der-Loeff[2], P. Masqué[3], G. Henderson[4]

^{236}U in seawater is dominantly of anthropogenic origin and has potential to become a new tracer of ocean circulation [1]. In this study, the first dataset of ^{236}U in Arctic Ocean waters is presented (together with ^{129}I), with the aim of mapping the distribution of ^{236}U and ^{129}I in the Arctic Ocean and interpreting the data in context of the different sources of radionuclides present in the Arctic Ocean.

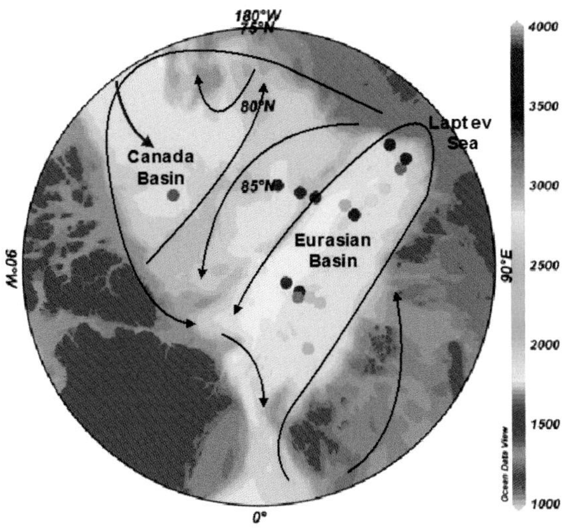

Fig. 1: *$^{236}U/^{238}U$ ($\cdot 10^{-12}$) atom ratio in surface waters of the Arctic Ocean. Black and blue arrows represent Atlantic Waters and Pacific Waters, respectively.*

A total of 20 surface samples were collected during two GEOTRACES expeditions in 2011/2012 (Polarstern ARK XVI/3 and ARK XVII/3) and analyzed for ^{236}U, $^{236}U/^{238}U$ and ^{129}I.

$^{236}U/^{238}U$ ratios (Fig. 1) in surface samples cover a range between 2000 and $4000 \cdot 10^{-12}$ in the Eurasian Basin. Lower values are observed in the Canada Basin (down to $1200 \cdot 10^{-12}$). ^{129}I concentrations in surface samples display even a broader range (between $10 \cdot 10^{7}$ and $800 \cdot 10^{7}$ at$\cdot$kg^{-1}), with the lowest value found in the Canada Basin, as for ^{236}U (Fig. 2).

Both ^{236}U and ^{129}I vary consistently with the different water masses found in the Arctic Ocean. Atlantic waters are characterized by high $^{236}U/^{238}U$ ratios and ^{129}I concentrations, due to releases from Sellafield and La Hague [2]. Pacific waters have lower ^{236}U and ^{129}I values, representing the global fallout signature [3].

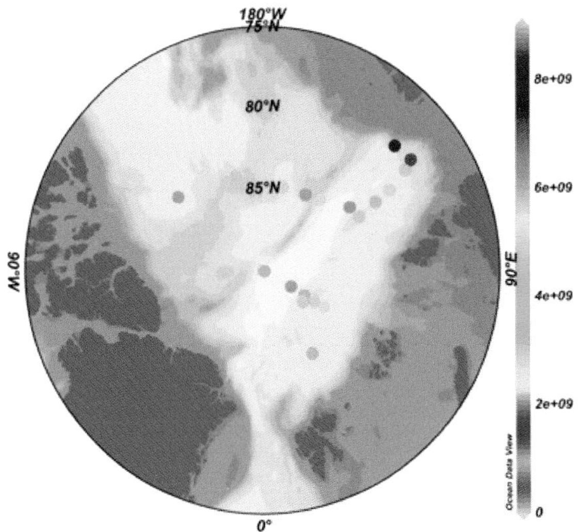

Fig 2: *^{129}I concentrations in surface waters of the Arctic Ocean.*

River input emerges as another potential source of ^{129}I and ^{236}U. This is observed in surface seawater samples collected close to the Laptev Sea, were river influence from Ob and Lena Rivers can be significant (up to 20%).

[1] N. Casacuberta et al., Geochim Cosmochim. Acta. 133 (2014) 34

[2] M. Christl et al., Nucl. Instr. & Meth. B 294 (2013) 530

[3] Sakaguchi et al., Earth and Plenet Sci Letters 333-334 (2012) 165

[1] *Institut für Radioökologie und Strahlenschutz, Germany*

[2] *Alfred Wegener Institute, Germany*

[3] *Universitat Autònoma de Barcelona, Spain*

[4] *Oxford University, United Kingdom*

DISTRIBUTION OF ^{236}U IN THE NORTH SEA (2009/2010)

Tracing the releases of nuclear reprocessing with ^{236}U

M. Christl, J. Lachner, N. Casacuberta, C. Vockenhuber, J. Herrmann[1], R. Michel[2],

The distribution of ^{236}U in the North Sea (Fig. 1) was investigated in two subsequent years (2009/2010). The main sources of ^{236}U in this region are the authorized liquid releases of uranium from the two nuclear reprocessing facilities Sellafield (GB) and La Hague (F). According to regional hydrography, the artificial tracer signals enter the North Sea where they homogenize and eventually leave the region via the North East towards the Arctic Ocean.

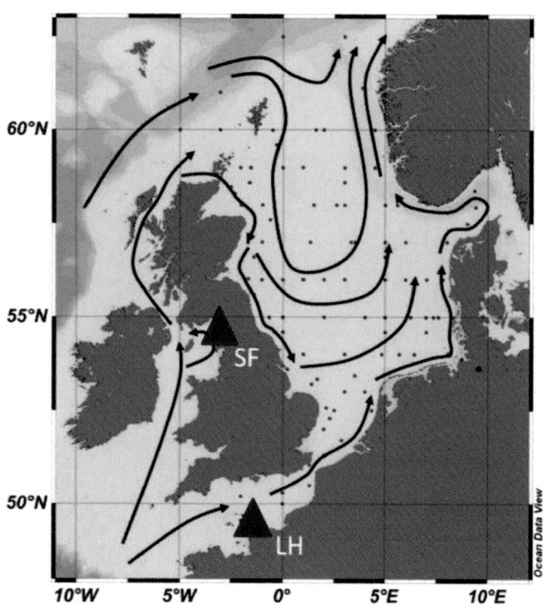

Fig. 1: Map of the North Sea region showing the *sampling locations in 2009 and 2010 (blue dots), the main surface currents (black arrows), and the location (red triangles) of Sellafield (SF) and La Hague (LH).*

Our results represent the first comprehensive dataset of ^{236}U/^{238}U in the North Sea region [1]. The combined results of two cruises carried out by BSH, Hamburg in 2009 and 2010 (Fig. 2) show a very consistent picture: The main signal with ^{236}U/^{238}U ratios at the order of several times 10^{-8} is found downstream of La Hague very close to the French, German, and Danish coasts.

A less pronounced signal of ^{236}U/^{238}U is also found in the Scottish coastal waters that very likely indicates ^{236}U releases from Sellafield.

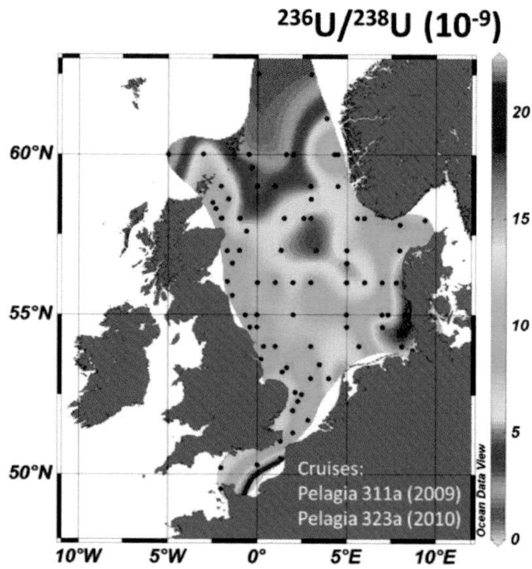

Fig. 2: *The distribution of ^{236}U/^{238}U (10^{-9}) in the North Sea in 2009 and 2010.*

The ^{236}U/^{238}U ratios in the North Sea are significantly larger than values of about 10^{-9} measured at the surface of the North Atlantic Ocean (mainly influenced by global fallout) [2]. This indicates that nuclear reprocessing facilities are the dominant source for the local input of ^{236}U in this region. We are currently investigating the suitability of ^{236}U (in combination with ^{129}I) as a new tool to determine transit times of Atlantic waters in the Arctic Ocean.

[1] M. Christl et al., Nucl. Instr. & Meth. B 294 (2013) 530

[2] N. Casacuberta et al., Geochim. Cosmochim. Acta 133 (2014) 34

[1] *BSH, Hamburg, Germany*

[2] *IRS Hannover, Germany*

MAPPING OF ^{236}U AND ^{129}I IN THE MEDITERRANEAN SEA

Tracing of water masses and sources of ^{236}U and ^{129}I

M. Castrillejo[1], N. Casacuberta, M. Chirstl, C. Vockenhuber, P. Masqué[1] and J. Garcia-Orellana[1]

The Mediterranean Sea is a semi-enclosed sea with very limited interaction with the open ocean. It has a rapid circulation time-scale and is heavily influenced by human activity. Anthropogenic radionuclides (hereinafter AR) were introduced to the marine environment since the 1950's through: i) nuclear weapon testing, ii) accidents such as Chernobyl and iii) releases from nuclear facilities. The basin-scale distribution of dissolved AR in the Mediterranean Sea has yet been poorly studied.

We produced the first comprehensive data set of ^{236}U and ^{129}I in the Mediterranean Sea from 10 full-depth profiles covering its main basins (Fig. 1). All analyses were performed on the compact AMS system TANDY at ETH Zürich.

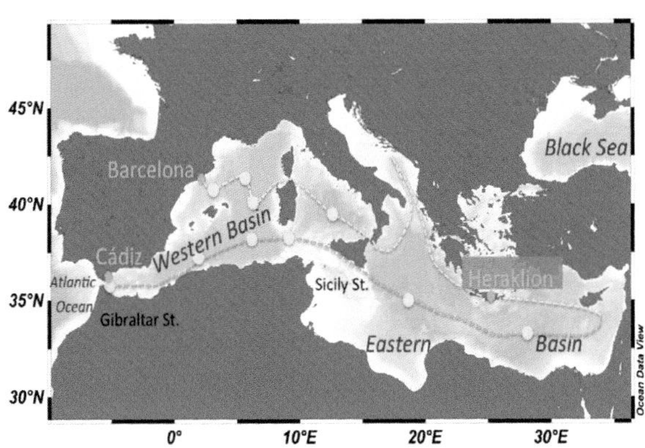

***Fig. 1:** Cruise track of the MedSeA/Geotraces GA-04-S cruise in May 2013. Yellow dots indicate sampled locations. Dashed lines show Leg 1 (green) and Leg 2 (yellow).*

^{236}U (Fig. 2) was recorded along the full water column with high ^{236}U/^{238}U atom ratios also in bottom waters ($\sim 1400 \cdot 10^{-12}$), reflecting the short mixing time-scales in the Mediterranean Sea. This is in contrast to the deep Atlantic Ocean, where ^{236}U/^{238}U ratios in Antarctic Bottom Waters can be lower than $100 \cdot 10^{-12}$ [1]. At surface, Atlantic Waters with low ^{236}U/^{238}U atom ratios ($1000 \cdot 10^{-12}$) enter through the Strait of Gibraltar. In the Eastern Basin, Levantine Intermediate Water (LIW) is formed with the highest values of this study ($2000 \cdot 10^{-12}$). At intermediate depths (1000-2000 m), low values associate to older water masses reaching $700 \cdot 10^{-12}$ in the Levantine Basin. At the bottom, ^{236}U/^{238}U ratios increase again due to the formation of (well-oxygenated) deep waters.

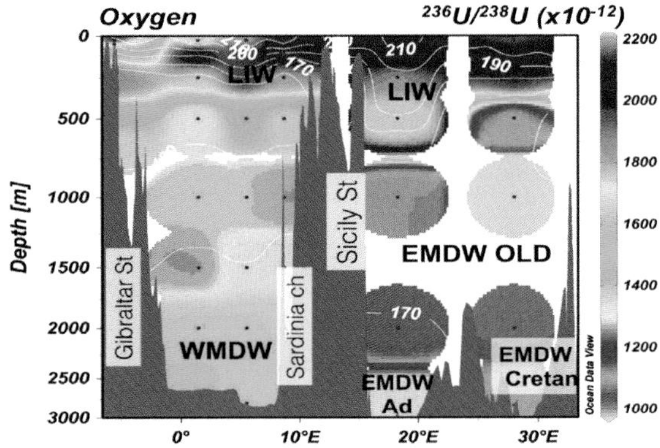

***Fig. 2:** Distribution of ^{236}U/^{238}U from west to east along Leg 1, from Gibraltar Strait (left) to the Levantine Basin (right).*

^{129}I (not shown) followed the distribution of ^{236}U and its concentrations ranged from 4 to $14 \cdot 10^{7}$ atom·kg^{-1}, leading to ^{129}I/^{236}U ratios between 4 and 12. These ratios point to the importance of other sources than global fallout. Other AR (^{137}Cs, ^{239,240}Pu and ^{90}Sr) will be analyzed in these samples. They will help identifying the sources of AR to the Mediterranean Sea and using them as tracers of circulation and particle dynamics.

[1] Casacuberta et al., Geo. Et Cosm. Acta 133 (2014) 34

---

1 Inst. de Ciència i Tecnologia Ambientals & Dep. de Física, Universitat Autònoma de Barcelona, Spain

SEPARATION OF ^{129}I IN SEAWATER SAMPLES

Setting up the ^{129}I radiochemistry at LIP/EAWAG

N. Casacuberta, M. Christl, C. Vockenhuber

^{129}I ($T_{1/2}$=15.7 Ma) is a powerful geochemical tracer in a wide range of natural reservoirs. One of the main applications is its use as an oceanographic tracer. The sources of anthropogenic ^{129}I in the marine environment are: i) nuclear weapons testing, ii) nuclear accidents and iii) continuous releases from nuclear reprocessing facilities, such as Sellafield (UK) and La Hague (France) [1].

During the year 2014, radiochemistry of ^{129}I in seawater samples has been set up at Laboratory of Ion Beam Physics (particularly in laboratories at the EAWAG) (Fig. 1). Samples are processed following the method described in [1].

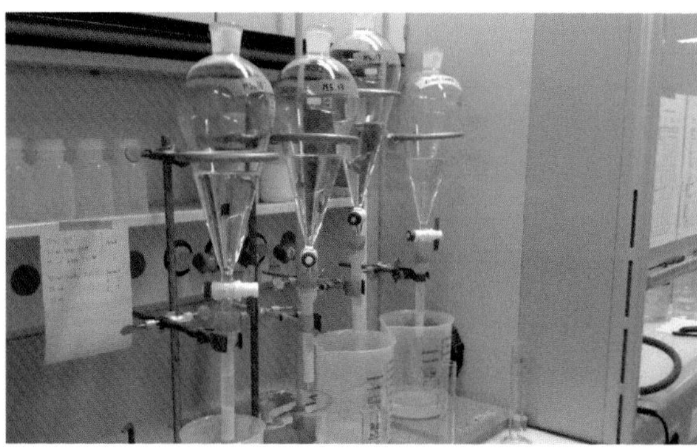

Fig. 1: *Chemistry setup in EAWAG labs.*

Briefly, Woodward iodine is added to the pre-weighed sample and all iodine species were oxidized with $Ca(ClO)_2$ to iodate. After 15 minutes, iodate species are reduced with NH_3OHCl and $NaHSO_3$ to iodide. After 45 minutes, the pH was raised to 5-6 before going through the separation step. This step consisted of an ion exchange separation using a BioRad ® 1x8 analytical grade resin. Resins are pre-conditioned with 0.5 M KNO_3 in order to increase the selectivity of the ion exchange resin. After the sample has gone through the resin, and the ion exchange columns rinsed with 0.5 M KNO_3, the iodine is eluted with concentrated potassium nitrate solution (2.25 M). Finally, iodine was precipitated as AgI for AMS measurement at ETH AMS TANDY [2].

The optimal ^{129}I/^{127}I ratio analyzed in the AMS TANDY should be in the range of 10^{-10} and 10^{-12}. Therefore, the amount of ^{127}I added to each sample and the volume of each sample to be analyzed, will depend on the expected ^{129}I concentration in seawater. In the Arctic and Atlantic Oceans, concentrations between 10^7 and 10^9 at·kg^{-1} are foreseen, due to the inputs from global fallout and nuclear reprocessing releases. Therefore, ideal sample volumes and amount of spike added is detailed in Table 1.

Expected ^{129}I in seawater [at·kg^{-1}]	Sample volume [L]	Added ^{127}I [mg]
10^7	0.450	1.5
10^8	0.200	3
10^9	0.150	3

Tab. 1: Optimal sample volumes and ^{127}I stable iodine added to each sample, according to the expected ^{129}I concentration in seawater.

A quality control has been performed to validate this method. We processed the certified reference material (IAEA-418) consisting of surface water collected in 2001 in the Mediterranean Sea and with a certified value of (2.3±0.2)×10^8 at·kg^{-1}. Four replicates were prepared and we obtained a final ^{129}I concentration of (2.6±0.1)×10^8 at·kg^{-1}.

In 2014, a total number of 100 samples were processed and measured at ETH Zurich following this method and more than 200 are expected for 2015-2016.

[1] R. Michel et al. Sci. Tot. Env. 419 (2012) 151
[2] C. Vockenhuber et al., AMS-13 proceedings

^{129}IODINE IN ENVIROMENTAL SAMPLES

Inventories, input and transport of iodine isotopes in Germany

A. Daraoui[1], M. Schwinger[1], M. Gorny[1], B. Riebe[1], C. Walther[1], C. Vockenhuber, H-A. Synal

The main source of anthropogenic ^{129}I in the environment is the release from European nuclear reprocessing plants of Sellafield (UK) and La Hague (F). Due to the anthropogenic input of ^{129}I, the stable ^{127}I and the long-lived ^{129}I exhibit a massive disequilibrium in all environmental compartments in Western Europe. Many of the ecological pathways of iodine are still unknown [1]. The main goal of this study is to investigate atmospheric input of ^{129}I, and its inventories in the pedosphere, together with the output of ^{129}I by rivers in Germany.

In this study, concentrations for ^{127}I and ^{129}I were determined in a number of samples by ICP-MS and low-energy AMS, respectively, and from that ^{129}I/^{127}I ratios were calculated in aerosol filter samples (n=4), precipitation samples (n=10), river water (n=15) and different soil samples (n=30) collected between 2011 and 2013 [2].

The 4 types of samples show distinct ^{129}I/^{127}I ratios with respect to ^{127}I concentrations (Fig. 1). The ranges for ^{127}I and ^{129}I concentrations and ^{129}I/^{127}I ratios in aerosol samples are 0.5-9.5 ng m^{-3}, 0.01-0.89 fg m^{-3}, and 8 x10^{-9}-6.8 x10^{-7}, respectively. In rain samples, they range from 0.6 to 60 µg kg^{-1} for ^{127}I, from 23 to 6155 fg kg^{-1} for ^{129}I, and from 1.4 x10^{-8} to 9.87 x10^{-7} for ^{129}I/^{127}I. In river water the results varied from 2 to 22 µg kg^{-1} for ^{127}I, from 19 to 810 fg kg^{-1} for ^{129}I, and from 0.3 x10^{-8} to 13 x10^{-8} for ^{129}I/^{127}I. The corresponding values for soil samples are 1-10 µg g^{-1}, 2-650 fg g^{-1}, and 9.6x10^{-10}-1.2x10^{-7}, respectively. Independent of the sample type concentrations of ^{129}I decreased from North to South and from West to East, respectively.

Generally, iodine evaporates from seawater, enters the atmosphere and is transported to the continent. In Germany, the ^{127}I concentrations are very uniform in all samples and do not influence the ^{129}I/^{127}I isotopic ratios, which are determined exclusively by the admixture of anthropogenic ^{129}I (Fig. 1). This clearly shows the disequilibrium of iodine isotopes in Germany.

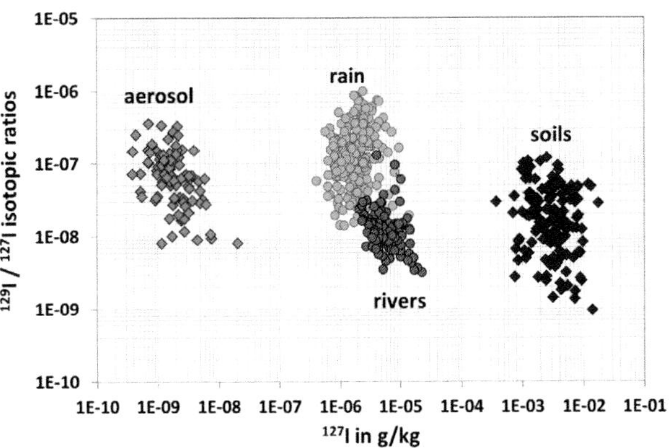

Fig. 1: *^{129}I/^{127}I isotopic ratios of all samples versus ^{127}I concentration.*

Besides the natural sources of ^{129}I, discharges from the mentioned reprocessing plants can be considered as main additional anthropogenic sources. The results from this project confirm the suggestion by Michel et al. [1] and show that the atmospheric and marine discharges from Sellafield and La Hague do significantly influence the fallout of ^{129}I in Germany, with the marine discharges being dominant.

[1] R. Michel et al., Sci. Tot. Environ. 151-169 (2012) 419

[2] B. Riebe et al., Abschlussbericht zum BMBF-Projekt 02 NUK 015D (2014) 159

---

[1] *Radioecology and Radiation Protection, University of Hannover, Germany*

ATMOSPHERIC INPUT OF ^{129}IODINE AT THE ZUGSPITZE

Determination of ^{129}I at high altitude Alp station Zugspitze in Germany

A. Daraoui[1], M. Gorny[1], B. Riebe[1], C. Walther[1], C. Vockenhuber, H-A. Synal

Atmospheric releases of ^{129}I from the European reprocessing plants in Sellafield (UK) and La Hague (F) can be transported by air masses over large distances. The goal of this study is to analyze the origin of ^{129}I in the Zugspitze area. Iodine concentrations were determined in snow and aerosol samples collected at the Alp Station Zugspitze at an altitude of 2650 m and 2962 m between 2012 and 2014. Iodine was separated from matrix elements by using an anion exchange method for snow samples and solvent extraction for aerosol filters. ^{127}I and ^{129}I were determined by ICP-MS and by AMS, respectively.

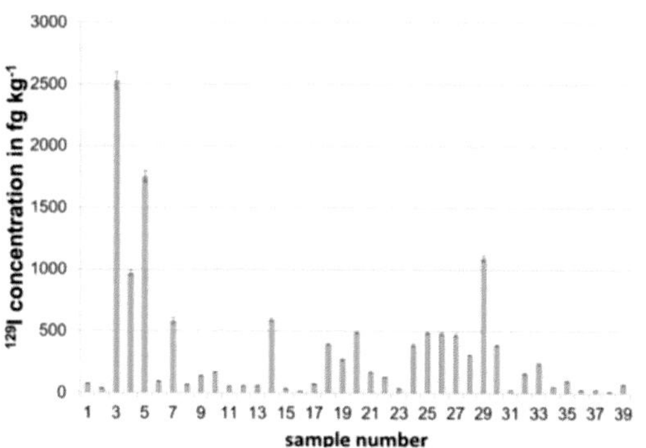

Fig. 1: *^{129}I concentration in snow samples at Zugspitze Schneefernerhaus station (2650 m).*

Concentrations of total ^{127}I in snow at 2650 m varied from 0.28 to 3.02 ng g^{-1} and for ^{129}I between 18 and 2529 fg kg^{-1} (Fig. 1). ^{129}I/^{127}I isotopic ratios ranged from 1.4x10^{-8} to 9.87x10^{-7}. The results indicate a larger variation in the concentrations of ^{129}I than for ^{127}I, which is due to a wide variation in the origin of air masses. High ^{129}I concentrations were measured on the 8th and 9th March 2012, and on the 15th May 2014, respectively. Analysis of the HYSPLIT Model of the backward trajectories in this period confirms that the air mass pathways are responsible for high ^{129}I concentrations, originating from the European reprocessing plants, and from the North Sea (Fig. 2).

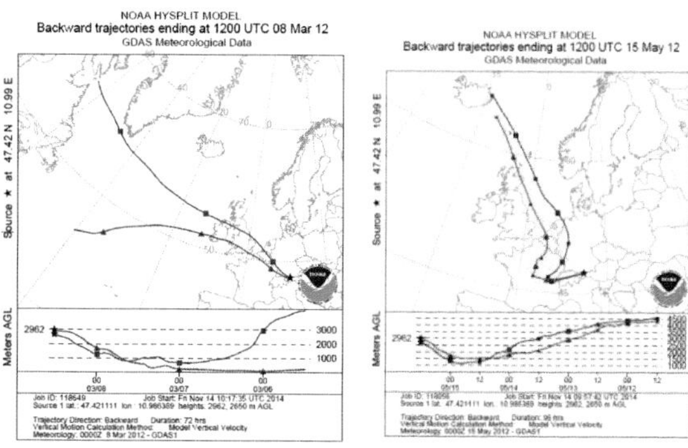

Fig. 2: *NOAA HYSPLIT examples of backward trajectories on 8.3.2012 and 15.5.2014.*

For ^{129}I concentrations in aerosol samples only small differences between sampling at 2650 m and 2962 m could be detected. The results varied between 0.002 and 0.035 fg m^{-3} (Fig. 3). The highest values were found in the period from 18th to 25th of August 2014. At an altitude of 2650 m we observed an ^{129}I concentration, which was 15 % higher than the one at 2962 m. First calculations indicate that wet deposition of ^{129}I at Zugspitze is more important than dry deposition.

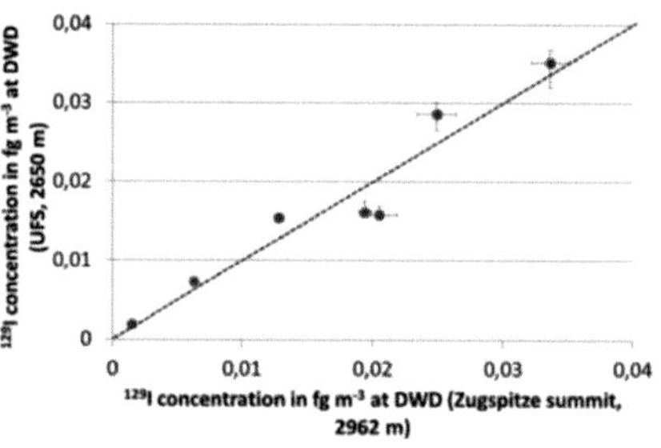

Fig. 3: *Correlation of ^{129}I concentrations at different sampling stations.*

---

[1] *Radioecology and Radiation Protection, University of Hannover, Germany*

$^{239/240}$Pu AND ^{236}U FROM EASTERN TIEN SHAN, CHINA

Chronology of actinide deposition at the Miaoergou glacier

H.W. Gäggeler[1], S. Hou[2], Ch. Wang[2], H. Pang[2], Y. Liu[2], M. Christl, H.-A. Synal

A 57.6 m long ice core was drilled in 2005 to bedrock at the eastern-most glacier of the Tien Shan mountains ranging from Kirgistan to China (43°03'19"N; 94°19'21"E; 4512 m asl) [1]. The core ranges back to about 1850 [2]. This site is interesting since it is a possible collecting location of nuclear debris from the Chinese (Lop Nor) and/or former Soviet (Semipalatinsk) test sites. Indeed, using β–counting, increased activity was observed between 7 m and 12 m depth which was assigned to the nuclear weapons fallout peak in 1962/63 and tentatively to Lop Nor [1].

Isotope ratios between ^{239}Pu and ^{240}Pu are a fingerprint of the emission source and can be measured very accurately with AMS even from ultra-trace amounts of plutonium stored in glacier ice cores. In addition, recently also ^{236}U has become a widely used tracer for environmental studies but its deposition rate chronology is still poorly known.

^{239}Pu, ^{240}Pu and ^{236}U were determined for the ice core section between 8 m and 15 m depth. This section corresponds to the period between about 1940 and 1970 [2].

The measured concentration profile showed a pronounced double peak with the following tentative assignment: higher concentrations during the post-moratorium period (maximum at about 1963) compared to the pre-moratorium period (maximum at about 1958) (see Fig. 1 for ^{239}Pu). The ^{240}Pu to ^{239}Pu ratios varied between 0.18 and 0.21, except for the two lowermost samples. These values are typical for global fallout and do not yield evidence for local sources from Lop Nor or Semipalatinsk.

The ratios (at/at) for ^{236}U to ^{239}Pu were between 0.15 and 0.52 (average 0.32) with rather large uncertainties due to some analytical problems in the separation of plutonium. These values agree with recently measured values in ice cores from Antarctica that range between 0.18 and 1.4 [3]. Some measured values in sea and river water samples were, however, significantly higher; they ranged from 1.0 to 12.0 [4]. Still, the deposition rates for ^{236}U on a global scale are poorly known and need further study.

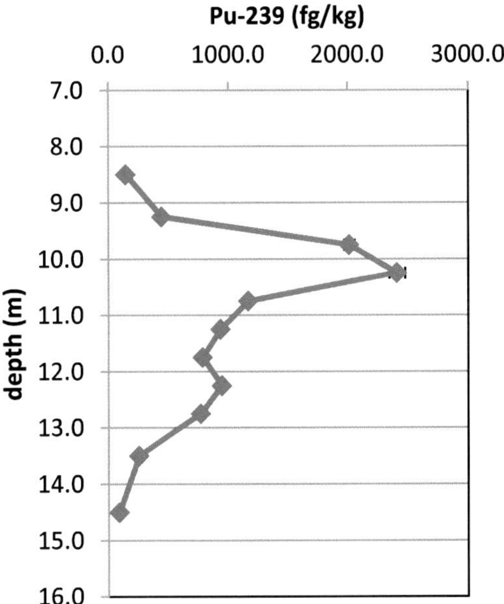

Fig. 1: *^{239}Pu concentration profile along part of the ice core. The higher peak at about 10 m is attributed to 1963 and the second, minor peak at 12 m to 1958, respectively.*

[1] Y. Liu et al., J. Geophys. Res., 116(D12), (2011) D12307

[2] Ch. Wang et al., Ann. Glaciol., 55(66), (2014) 105

[3] C.C. Wendel et al., Sci. Tot. Environ. 461, (2013) 734

[4] R. Eigl et al., J. Env. Radioact. 116, (2013) 54

[1] Paul Scherrer Institut, Villigen, Switzerland

[2] Nanjing University, China

PU BIOAVAILABILITY IN FRESHWATER ENVIRONMENTS

In-situ Pu measurements using diffusive gradients in thin films

R. Cusnir[1], P. Steinmann[2], M. Christl, F. Bochud[1], P. Froidevaux[1]

The fate of plutonium in the environment is determined to a large extent by its redox speciation and its interactions with naturally occurring colloids and organic matter (NOM). Studies on the speciation of Pu in aquatic environments require sampling of large volumes of water, as well as ultrafiltration techniques.

Here, we applied the technique of diffusive gradients in thin films (DGT) for in-situ Pu bioavailability measurements in the Venoge spring and Noiraigue Bied brook of the Swiss Jura Mountains.

Fig. 1: *DGT samplers with 105 cm² surface (left) for Pu bioavailability measurements exposed in the Venoge spring (right).*

A DGT sampler device is an assembly of three layers: the Chelex resin binding phase, the polyacrylamide (PAM) gel diffusive layer of a given thickness and a filter membrane protecting the gel (Fig. 1). Free and labile Pu species diffusing through the PAM gel layer are accumulated in the binding phase which is further analyzed using the ETHZ TANDY AMS system tuned for the analysis of actinides. The amount of Pu accumulated in the Chelex resin allows calculating the concentration of bioavailable species in the bulk water, using the diffusion coefficient of Pu determined previously [1].

The concentration of ^{239}Pu measured by DGTs (Tab. 1) was found in the same range as previously reported for the Venoge spring water (5-10 μBq L^{-1} measured in 200 L by alpha-spectrometry) [2].

Sample type	Number of samples	^{239}Pu, μBq L^{-1}
Venoge bulk	4	1.92
DGT 0.39 mm	4	1.19
DGT 0.78 mm	4	1.67
Noiraigue bulk	3	3.05
DGT 0.39 mm	3	0.68
DGT 0.78 mm	3	1.57

Tab. 1: *^{239}Pu activity concentrations measured by AMS in water samples and in the DGTs.*

DGTs indicate that about 70% of ^{239}Pu in the Venoge spring is bioavailable, however this value is different (30 and 60%) for the Noiraigue organic-rich water when measured with DGTs of different gel thicknesses. This important variation can be explained by considering kinetic factors of ^{239}Pu interactions with NOM molecules.

[1] R. Cusnir et al., Environ. Sci. Technol. 48 (2014) 10829

[2] P. Froidevaux et al., Environ. Sci. Technol. 44 (2010) 8479

[1] *Institute of Radiation Physics, CHUV, UNIL*

[2] *Federal Office of Public Health, Bern*

BIOASSAY METHODS FOR ACTINIDES BY COMPACT AMS

Ultra-sensitive detection of Pu, Np, Am, Cm and Cf isotopes in urine

X. Dai[1], M. Christl, S. Kramer-Tremblay[1], H.-A. Synal

Measurements of ultra-low levels of actinides in urine samples are often required for dose assessment when incidental exposure to these highly toxic radionuclides occurs. While the analyses of U, Th and Pu isotopes in urine samples using mass spectrometry have already become routine at the Chalk River Laboratories (CRL), Canada, more sensitive urine bioassay methods for minor actinides such as Np, Am, Cm and Cf radionuclides have yet to be developed. Accelerator mass spectrometry (AMS), particularly the compact systems, have evolved over the past years as one of the most sensitive, selective and robust techniques for actinide analyses [1, 2]. Employment of the AMS technique can reduce the demands on sample preparation chemistry and increase sample analysis throughput, due to its excellent sensitivity, high rejection of interferences and low susceptibility to adverse sample matrices.

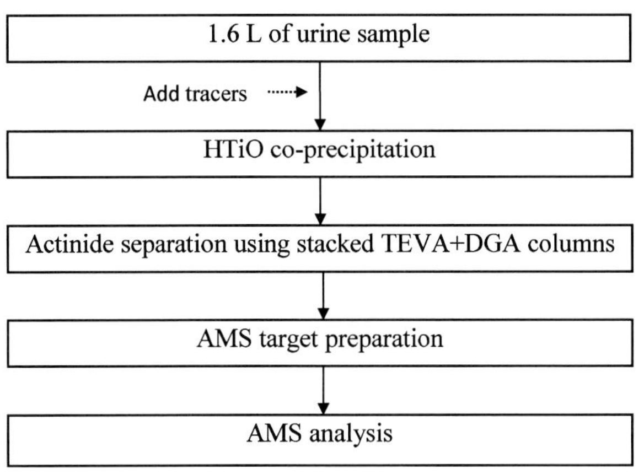

***Fig. 1:** Flow diagram of urine bioassay procedure for actinides by AMS.*

Initial studies were performed to explore and demonstrate the analytical capability of AMS for actinide urine bioassay. Blanks and urine samples spiked with atto- to femtogram levels of Np, Pu, Am, Cm and Cf isotopes were prepared, processed through the bioassay procedure (Fig. 1), and measured using the Tandy AMS system at ETH. The results agreed well with the expected spike values at fg/L levels or even less (see Fig. 2 for Cm-244, the results for Np, Pu, Am, Cf and other Cm isotopes are not shown), demonstrating the utility of compact AMS for the ultra-sensitive analysis of intermediate- and long-lived actinides in large volume of urine bioassay samples.

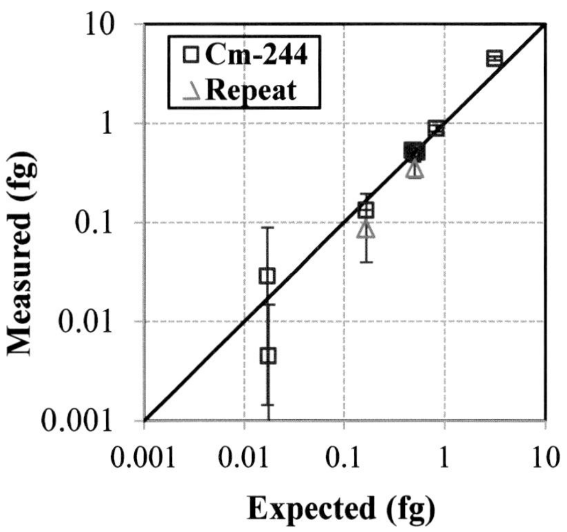

***Fig. 2:** Measured vs. expected ^{244}Cm in the spiked 1.6 L urine samples.*

The detection limits of the present bioassay method were estimated to be in the range of several ag/L for urine samples. This would provide sufficient sensitivity to meet the requirements of most bioassay scenarios.

[1] X. Dai, M. Christl et al., J. Anal. Atom. Spectrom., 27 (2012) 126

[2] M. Christl, X. Dai et al., Nucl. Instr. and Meth. B 331 (2014) 225

[1]Chalk River Laboratories, Canadian Nuclear Laboratories, Canada

PREPARATION OF A MULTI-ISOTOPE PU AMS STANDARD

Preliminary results of a first inter-lab comparison

B.-A. Dittmann[1], T. J. Dunai[1], C. Feuerstein[2], L. K. Fifield[3], M. Christl

Fallout radionuclides have been used for many years as tracers for soil and sediment transport studies. One of the most commonly used radioisotopes is fallout ^{239}Pu. The detection of fallout derived plutonium at ultra-trace levels is commonly performed by accelerator mass spectrometry (AMS). However, stocks of existing Pu-standard solutions are declining. Therefore, the motivation of this work was to establish a new multi-isotopic plutonium standard (containing ^{239,240,242,244}Pu) for isotopic ratio measurements.

The certified reference materials used for the preparation of the new multi-isotope plutonium standard were obtained from JRC IRMM. The isotopic composition of the materials is shown in Tab. 1.

Description	Mass fraction (· 100)			
	Pu-239	Pu-240	Pu-242	Pu-244
CRM Pu-239	**97.763**	2.202	0.007	0
CRM Pu-240	0.614	**98.963**	0.054	0
CRM Pu-242	0.239	8.528	**87.866**	0.019
CRM Pu-244	0.033	0.666	1.322	**97.908**

Tab. 1: *Isotopic composition of the used CRMs.*

The original solutions were mixed and diluted several times to obtain sufficient material that allows the preparation of ~100,000 AMS cathodes, assuming that each cathode will be loaded with approximately 5 pg of each of the abundant plutonium isotopes. The final solution contains the following amount of plutonium isotopes: N(^{239}Pu) = $(1.164 \pm 0.001)\cdot 10^{13}$ at/g, N(^{240}Pu) = $(1.173 \pm 0.001)\cdot 10^{13}$ at/g, N(^{242}Pu) = $(1.108 \pm 0.003)\cdot 10^{13}$ at/g and N(^{244}Pu) = $(1.150 \pm 0.003)\cdot 10^{12}$ at/g. The quoted plutonium concentrations are calculated based on the certified values and the weights.

The AMS measurements were performed in a round robin exercise at three labs, ANU Canberra, ETH Zurich and CologneAMS.

Facility	Ratio [at/at] normalized to UKAEA			
	^{240}Pu/^{242}Pu	^{239}Pu/^{242}Pu	^{240}Pu/^{239}Pu	^{244}Pu/^{242}Pu
ETH	1.085 ± 0.028	1.054 ± 0.004	1.028 ± 0.026	0.107 ± 0.003
ANU	1.062 ± 0.007	1.094 ± 0.037	0.972 ± 0.028	0.107 ± 0.002
ColAMS	1.060 ± 0.018	1.047 ± 0.010	1.013 ± 0.011	0.102 ± 0.004
Nominal	1.050 ± 0.003	1.058 ± 0.003	1.007 ± 0.001	0.104 ± 0.004

Tab. 2: *Arithmetic means of individual ratios. The error is given as standard deviation (1σ).*

All AMS labs performed two measurement series. A comparison of the arithmetic mean values normalised to the UKAEA standard are in good agreement. All mean values are within the range of 1-3% as the individual results would suggest. The scatter between individual measurements was random, thus more measurements will improve the final uncertainties.

It is planned to include further AMS facilities in the round robin exercise and to continue the measurement of the new plutonium AMS standard. A verification of the nominal ratio by Multicollector-ICP-MS is intended.

[1] *Geology, University of Cologne, Germany*
[2] *Nuclear Physics, University of Cologne, Germany*
[3] *Research School of Physics, ANU, Australia*

LONG-LIVED HALOGEN NUCLIDES IN MEGAPIE

Radiochemical determination of ^{129}I and ^{36}Cl in the PSI MEGAPIE target

B. Hammer[1], A. Türler[1], D. Schumann[2], J. Neuhausen[2], V. Boutellier[2], M. Wohlmuther[2], C. Vockenhuber

The concentration of the long-lived nuclear reaction products ^{129}I and ^{36}Cl has been measured in samples from the MEGAPIE liquid metal spallation target at PSI. Samples from the bulk target material (lead-bismuth-eutectic, LBE), from the interface of the metal-free surface with the cover gas, from LBE/steel interfaces and from noble metal absorber foils installed in the cover gas system were analyzed. Bulk LBE samples contain only a minor part of the ^{129}I and ^{36}Cl detected, whereas most of these nuclides were found accumulated on the interfaces.

Visual inspection of the samples from the cover gas interface reveals a contamination of the LBE with solids of different appearance. According to earlier studies [1], these solids are enriched in radionuclides of electropositive elements that have separated from the liquid metal. The results of the present work indicate that the same solids are also enriched in ^{129}I and ^{36}Cl. Similarly, in samples from the LBE/steel interface that were found to be enriched in radionuclides of electropositive elements [1] the present work reveals also an enrichment of ^{129}I and ^{36}Cl. A possible explanation for the enrichment of electropositive elements in surface layers offered in [1] was that these elements form oxides from oxygen impurities present in the LBE or from the protective oxide layers of steels which are practically insoluble in the liquid metal and thus precipitate on the container walls and on the free surface of the liquid metal. For the elements I and Cl, a chemically plausible explanation for the enrichment in surfaces could be incorporation of halogenide ions into such precipitates.

In order to estimate the total amount of ^{129}I and ^{36}Cl present in the target from the analytical data, surface activities were estimated based on the geometry of the samples and extrapolated to the complete target inner surface. In Tab. 1 the presently available results for ^{129}I are summarized and compared with predictions (average of FLUKA and MCNPX calculations) of the total amount of this isotope produced in the MEGAPIE target [2]. The amount of ^{129}I estimated from the analytical data agrees reasonably well with the results of nuclear calculations of 8.6x10^3 Bq [2]. For ^{36}Cl, qualitatively similar results were obtained. However, the larger scatter of the data within the individual types of samples impedes reliable extrapolation to the complete target.

	measured [Bq]	% of predicted amount[c]
bulk	2.6x10^2	3
free surface[a]	1.2x10^1	<1
LBE/steel interface[b]	3.7x10^3	44
absorber	4.0x10^{-1}	<1
sum	4.0x10^3	47

Tab. 1: *Summary of ^{129}I activity in different sample types; [a]inner surface of LBE container 16 m^2; [b]LBE/cover gas interface approx. 300 cm^2; [c]predicted amount is the average of two calculations using different nuclear models (8.6x10^3 Bq).*

The work was funded by the EU projects ANDES and GETMAT in the frame of EURATOM FP7.

[1] B. Hammer et al., J. Nucl. Mater., 450 (2014) 278

[2] L. Zanini et al., PSI Report nr. 08-04 (2008)

---

[1] *Radio.- and Environmental Chemistry, PSI, Villigen and Chemistry and Biochemistry, University of Bern*

[2] *Radio.- and Environmental Chemistry, PSI, Villigen*

PRODUCTION OF ^{129}I AND ^{36}Cl IN LEAD

Completion of the dataset of long-lived radionuclides in the SINQ target

T. Lorenz[1], D. Schumann[1], R. Dressler[1], Y. Dai[2], M. Wohlmuther[1], D. Kiselev[1], A. Türler[1,3], C. Vockenhuber

We investigated the radionuclide inventory of a solid lead target SINQ that was in operation at PSI from March 2000 to December 2001. A total beam dose of 10.03 Ah was deposited inside an arrangement of more than 300 lead filled steel rods. Samples from different positions inside the target (Fig. 1) were analyzed for a set of 19 relevant long-lived radionuclides using γ-ray and α-particle spectrometry and AMS. This kind of data set is requested from the Swiss authorities to issue permission for conditioning and disposal of SINQ targets.

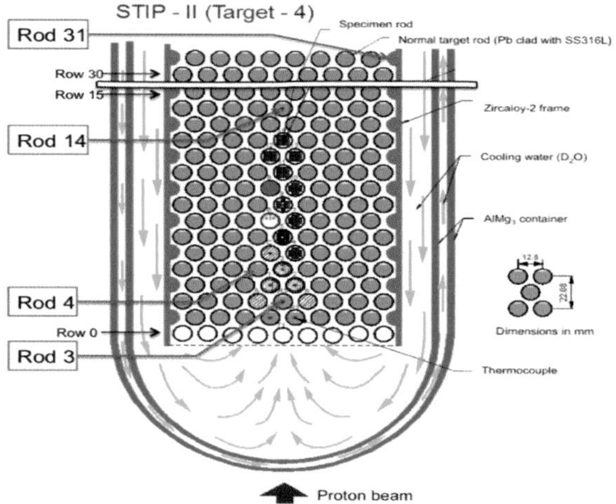

Fig. 1: *Scheme of the rod arrangement in Target 4. Sample rods are indicated.*

In previous works, the studies on the activity concentration of strong γ-emitters [1], Po-isotopes [2], ^{36}Cl and ^{129}I [3] and α-particle emitting lanthanides ^{146}Sm, ^{148}Gd and ^{150}Gd [4] in samples from Rod 3 were presented. Further analysis of samples along the beam axis provides a data basis to calculate the total radionuclide inventory and its spatial activity distribution. The integral radioactive inventory was calculated using the interpolation method of Kriging [5]. The total spatial distribution (radial and in direction of beam) is visualized in a two-dimensional contour plot (Fig. 2).

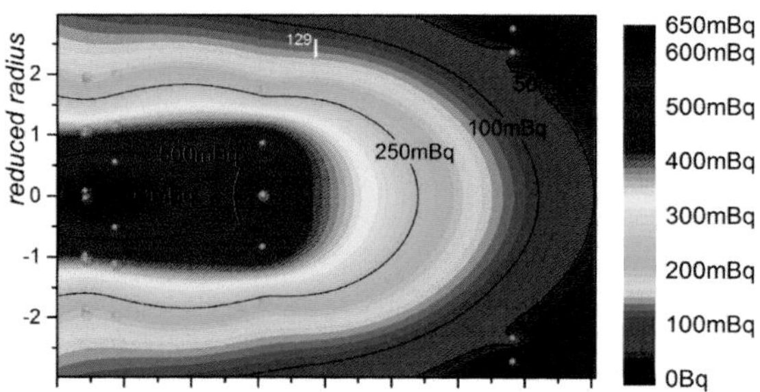

Fig. 2: *2D interpolation of ^{129}I using the Kriging algorithm. The data points are indicated in grey.*

The total activities, scaled to end of beam, were determined to be 4.43x10^3 Bq (±26%) for ^{129}I and 1.74x10^5 Bq (±29%) for ^{36}Cl. The results are in good agreement with theoretical predictions using two physics models (Fig. 3).

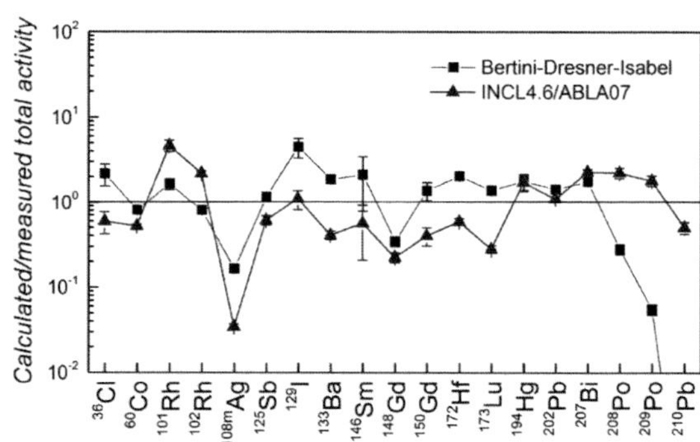

Fig. 3: *Ratio of calculated to measured data.*

[1] T. Lorenz et.al., Y. Dai, RCA 101 (2013) 661
[2] T. Lorenz et.al., NDS 119 (2013) 284
[3] T. Lorenz et.al., Ann. Rep. LCH (2012) 46
[4] T. Lorenz et.al., Ann. Rep. LCH (2013) 48
[5] G.M. Laslett, J. Am. Stat. Ass. 89 (1994) 391

[1] *Radio.- and Environmental Chemistry, PSI, Villigen*
[2] *Nuclear Materials, PSI, Villigen*
[3] *Chemistry and Biochemistry, University of Bern*

MATERIAL SCIENCES

MeV SIMS PROGRESS: INTRODUCING CHIMP

Status of the new MeV SIMS setup at the EN Tandem accelerator

M. Schulte-Borchers, M. Döbeli, A.M. Müller, M. George, H.-A. Synal

A new MeV SIMS (Secondary Ion Mass Spectrometry) setup based on a capillary microprobe was designed and constructed. The device was named with the acronym CHIMP: Capillary Heavy Ion Micro Probe. It combines the novel technique MeV SIMS [1] with proton induced X-ray emission (PIXE) and scanning transmission ion microscopy (STIM). Due to the specific placement of the analysis chamber in the high-energy beam line of the 6 MV EN Tandem accelerator with only electrostatic ion optical elements, it is possible to use very heavy and energetic primary beam particles.

Fig. 1: *CHIMP vacuum chamber. The MeV ion beam enters from right, the ToF drift tube protrudes to the left.*

The UHV chamber (Fig. 1) is equipped with a sample entry lock for a magazine accommodating five samples and a separate turbo pump for fast magazine changes. Each sample can be manually transferred from the magazine onto a piezo based XY rastering stage using a wobble stick. The stage has an imaging range of 15x15 mm^2.

Primary ions impinge on the sample surface under normal incidence. They can be collimated with a glass capillary to a spot size of a few micrometers. Under 45° to the primary beam secondary ions are extracted into a time of flight (ToF) spectrometer with a field-shaping cone and acceleration grids (Fig. 2). After a drift path of 0.5 m the ions are detected by an MCP detector.

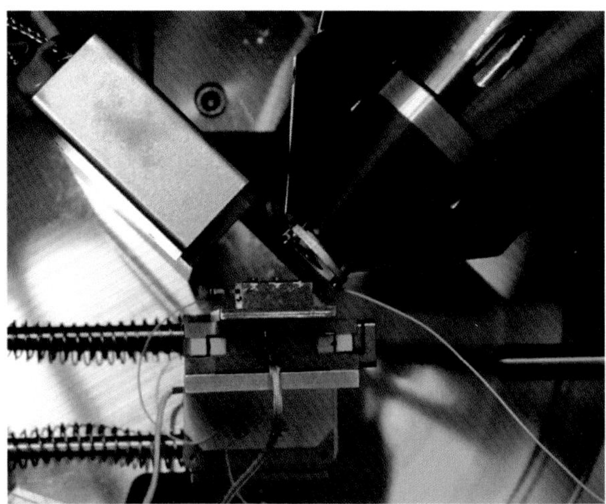

Fig. 2: *Close view of the target region showing the snout of the ToF spectrometer (top right).*

Besides secondary ions, secondary electrons and X-rays emitted from the sample surface can be detected with corresponding detectors in the chamber. A gas ionization chamber behind the sample allows to detect direct beam particles and provides the possibility to perform STIM measurements. The target region can be observed by a camera with a high magnification telephoto lens.

First test measurements have been conducted with promising results. They confirm the general functionality of the CHIMP system. Evaluation of ToF trigger modes and further optimization of the secondary ion and electron extraction parameters are in progress.

[1] B.N. Jones et al., Nucl. Instr. and Meth. B 268 (2010) 1714

PREFOCUSSING INTO THE CAPILLARY MICROPROBE

Improving the beam current using magnetic lenses

M. Schulte-Borchers, A. Bortis, M. Döbeli

The capillary microprobe at ETHZ has proved to be a useful tool for microanalysis and imaging of big or sensitive samples in air. It provides micro beams of diameters in the range of one to ten micrometers and is easy to setup which makes it a cheap and simple alternative to conventional microprobes using magnetic micro lenses. However, due to missing focusing properties [1] beam currents from capillaries are rather low in comparison. Prefocusing with a low quality micro lens can enhance the current density in the capillary and increase the output current while aberrations and astigmatisms are cut off by the capillary collimation.

Fig. 1: *Picture of the original micro beam setup at ETHZ, showing the quadrupole lenses.*

The existing magnetic micro lenses are a doublet of magnetic quadrupoles (Fig. 1) purchased in the 1970s from Auckland Nuclear Accessory Company and had been reported to focus ions down to 10-30 μm^2.

We performed now prefocusing experiments with capillaries of 0.7, 5.4 and 7.1 micrometer diameter. To evaluate the achieved beam current in air without influence of charges produced by ionization, PIXE yields obtained with the capillary setup have been compared to those produced by the same iron sample in the RBS chamber where the beam current can precisely be measured using the yield of backscattered ions. Careful corrections for the different geometry and absorption conditions have been applied.

Fig. 2: *Close-up on the external PIXE setup at the capillary, showing the iron sample (right) and X-ray detector in the background.*

As a result, in-air beam currents of the order of 10 pA have been achieved for the 5.4 and 7.1 micrometer capillaries. PIXE yields of several hundred counts per seconds were obtained from the iron sample. This is enough to enable fast single spot measurements and even PIXE imaging.

The smallest capillary yielded a current of almost 20 pA. This inconsistency points towards damaging of the capillary tip during the experiment. Since such an incident has been observed for the first time, the durability of especially very small capillaries at high current densities has to be further investigated.

[1] M.J. Simon, et al., Nucl. Instr. and Meth. B330 (2014) 11

FIRST TRITIUM TO DEUTERIUM RATIOS BY AMS AT LIP

Neutron flux measurement via tritium production in deuterium

H. Becker[1], K. Kirch[1,2], B. Lauss[2], D. Ries[1,2], M. Döbeli, L. Wacker

Two methods of measuring the cold neutron flux through the solid deuterium moderator of the ultra-cold neutron source at PSI have been explored. Both approaches are based on measuring the concentration of tritium produced by neutron activation of deuterium.

The tritium content of gas samples taken from the moderator is conventionally determined by liquid scintillation counting (LSC). For this purpose the gas has to be liquefied using a fuel cell. As an alternative technique, we analyzed aliquots of the same gas samples by AMS with the LIP MICADAS. Low-level mass spectrometry of hydrogen isotopes with conventional instruments is problematic due to interferences with H_2 and HD molecules.

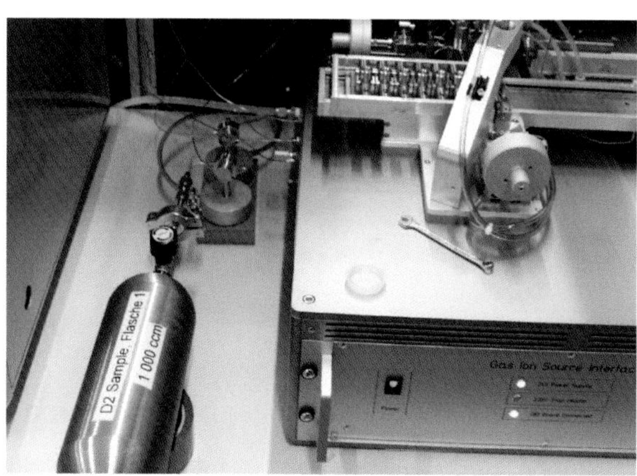

Fig. 1: *Deuterium sample bottle connected to the gas interface of the MICADAS negative ion source.*

The sample gas has been directly fed to the gas interface of the negative ion source. All three hydrogen isotopes have been analyzed. For protons and deuterium the ion current was measured behind the high energy magnet, while tritium ions had to be counted in the gas detector due to the small concentrations. Because of the big relative mass differences of hydrogen isotopes it is impossible to use the fast beam switching system. Therefore, the MICADAS magnets were manually adjusted to the mass of the three isotopes. The background count rate obtained with pure deuterium was subtracted from all measured tritium count rates. Fig. 2 shows a preliminary comparison of the tritium to deuterium ratios measured by the two techniques. The tritium concentration obtained by the LSC method tends to be higher than those by AMS. This can be due to isotopic effects during deuterium liquefaction or caused by transmission differences in the AMS system.

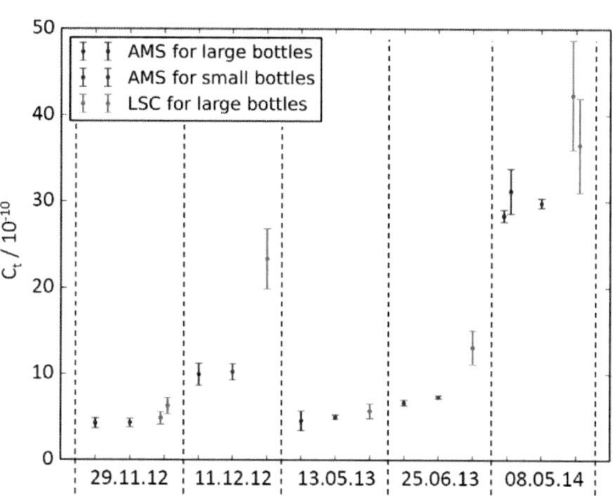

Fig. 2: *Preliminary results for the tritium to deuterium isotopic ratios. AMS and LSC values are compared for five different sampling dates.*

For more accurate results a suitable tritium to deuterium gas standard would be desirable. To obtain a more direct comparison of LSC and AMS results it was tried to feed water vapor from the liquefied samples to the gas ion source, but to date no sufficient ion currents have been obtained.

---

[1] *Particle Physics, ETH Zurich*
[2] *Paul Scherrer Institut, Villigen*

BERYLLIUM ENRICHMENT AT COPPER SURFACE

A tentative pathway to obtain ultra-cold neutron reflection

S. Parolo[1], F. Piegsa[1], K. Kirch[1], A.M. Müller, M. Döbeli

Ultra cold neutrons (UCN) move with a speed of less than 10 m/s. Therefore they can be guided and stored in traps. The surface of neutron guides and traps usually consists of a neutron reflecting material, such as beryllium or beryllium oxide.

Fig. 1: *Copper beryllium samples. The polished face and mounting holes are shown.*

The goal of this work was to produce a Be enriched surface by heat treatment of commercial beryllium copper (CuBe), a copper alloy with 0.5 - 3 wt% beryllium [1]. Polished plates of 2 wt% CuBe (Fig. 1) were annealed in three types of vacuum furnaces. Temperatures ranged from 500 to 825 °C and treatment times from 30 minutes to 65 hours. The Be concentration depth profiles were then analyzed by Heavy Ion ERDA using 13 MeV ^{127}I projectiles.

Fig. 2 shows a comparison of two Be profiles obtained at 500 and 750 °C with a treatment time of 2 hours. The Be/Cu ratio was divided by the value of untreated samples to obtain the net "Be gain". As expected, Be enrichment increases with annealing temperature and treatment time.

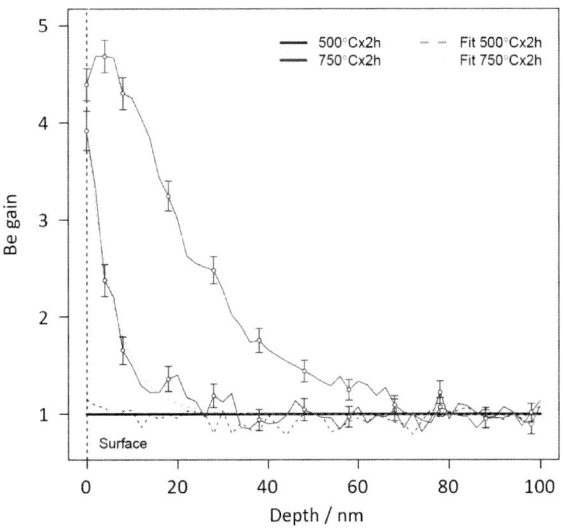

Fig. 2: *Measured depth profiles of the beryllium gain for two heat treated samples.*

For all utilized furnaces, a distinct enhancement of beryllium at the surface was observed. However, the maximum obtained concentration is not yet sufficient to obtain a suitable neutron reflecting surface. The efficiency of the Be segregation can possibly be improved by longer heat treatment at even higher temperatures, by exposure of the surface to ammonia or by oxygen beam assisted annealing.

[1] F. Watanabe, Journal of Vacuum Science and Technology A22 (2004) 181

---

[1] *Particle Physics, ETH Zurich*

GAS SCATTERING IN PULSED LASER DEPOSITION

Angular dependence of film composition measured by RBS and ERDA

A. Ojeda[1], C.W. Schneider[1], Th. Lippert[1], M. George, M. Döbeli

In Pulsed Laser Deposition (PLD) the transfer of material from the ablation target to the sample substrate is mediated by a dense propagating plasma plume which expands in vacuum or in a background gas of specific pressure. In general, a congruent thin film growth with the original stoichiometry of the target is not achieved. The ablation process itself, plasma propagation effects or scattering of plasma constituents in the background gas can be reasons for this [1].

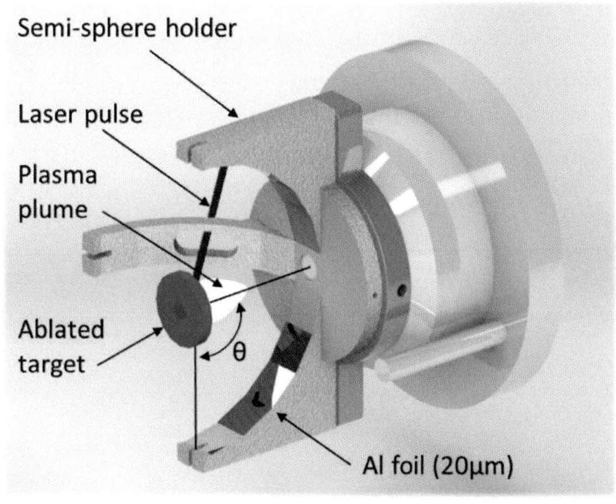

***Fig. 1:** Schematic design of the experimental set-up used for PLD deposition on catcher foils.*

The goal of the project was to investigate the role of scattering in the background gas and to set the experimental basis for model calculations. Film thickness and composition as a function of the angle towards the laser beam has therefore been measured for several target compositions and gas pressures by means of Al catcher foils (Fig. 1). For $BaTiO_3$ and $La_{0.4}Ca_{0.6}MnO_3$ the deposits have been analyzed by RBS. For $LiMnO_3$, which is of special interest due to the large Mn/Li mass ratio, the composition was determined by a combination of RBS and ERDA. Fig. 2 shows the angular distribution of the $BaTiO_3$ film thickness for two gas pressure values. The homogenizing effect of scattering in the background gas is distinct.

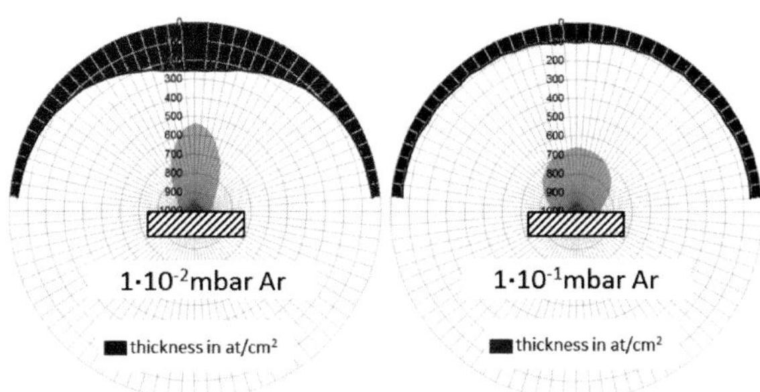

***Fig. 2:** Polar diagrams of $BaTiO_3$ film thickness as a function of angle towards the laser beam for two values of Ar background gas pressure.*

In Fig. 3 the Ca and Mn content normalized to a fixed value of La of 0.4 is shown for deposition of $La_{0.4}Ca_{0.6}MnO_3$ in 10^{-2} mbar of Ar background gas. Preferential scattering of the lighter elements in the gas can clearly be observed.

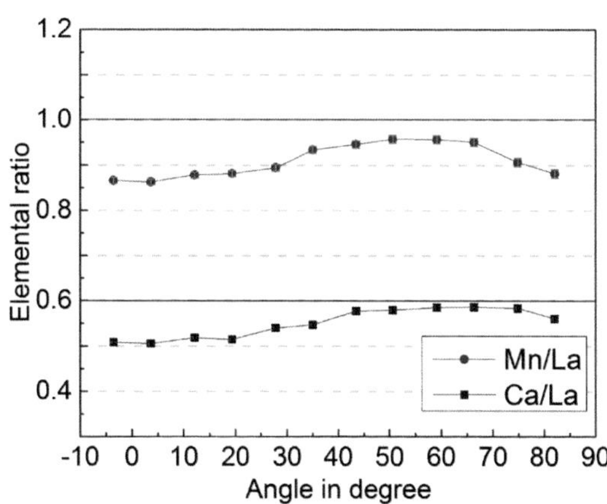

***Fig. 3:** Mn and Ca content as a function of angle for deposited $La_{0.4}Ca_{0.6}MnO_3$ in 10^{-2} mbar of Ar.*

[1] J. Chen et al., Appl. Phys. Lett. 105 (2014) 114104

[1] *Materials Group, PSI, Villigen*

PROCESS CONTROL OF SPUTTER DEPOSITED THIN FILMS

ERDA and RBS analysis of aluminum oxynitride layers

M. Fischer[1], M. Trant[1], K. Thorwarth[1], H.J. Hug[1], J. Patscheider[1], A.M. Müller, M. Döbeli

Thin films of aluminum oxynitride with varying contents of N and O were prepared at EMPA using reactive direct current (DC) magnetron sputter deposition (Fig. 1). In order to gain control over the process and for an appropriate correlation between film composition and physical properties, a detailed elemental analysis by ERDA and RBS has been undertaken.

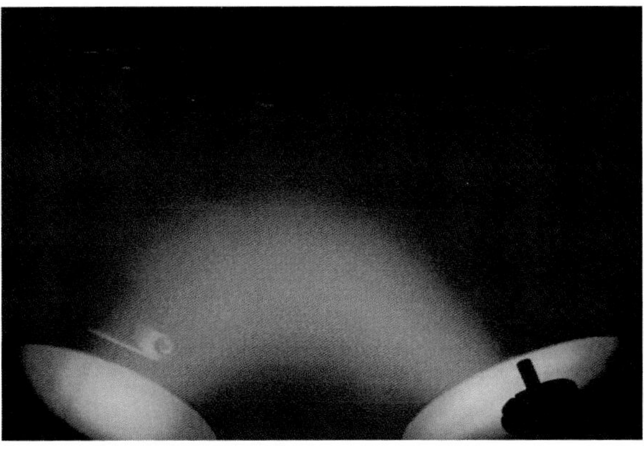

***Fig. 1:** View of the DC magnetron sputter deposition chamber with substrates to be coated at the top.*

To produce films with different N/O ratios the flow of the reactive gases N_2 and O_2 was varied at constant Ar working gas flow and power on the Al sputter target. After deposition, the composition of the material was determined by 2 MeV ^{4}He RBS and 13 MeV ^{127}I Heavy Ion ERDA. Fig. 2 shows an example for a batch of samples, for which the O_2 content in the N_2/O_2 mixture was raised from 0 to 10 %. The O content in the films of this batch rises quickly upon increasing O_2 flow. At 10% O_2 in the reactive gas the N incorporation almost completely ceases.

Analysis by ERDA and RBS provides the possibility to investigate the depth profile of the atomic composition. Such a profile is displayed in Fig. 3, which shows the O and N profiles of a sample grown under unstable conditions. Obviously, some N was incorporated in the film at the beginning of the deposition while the rest of the material consists of virtually pure Al_2O_3.

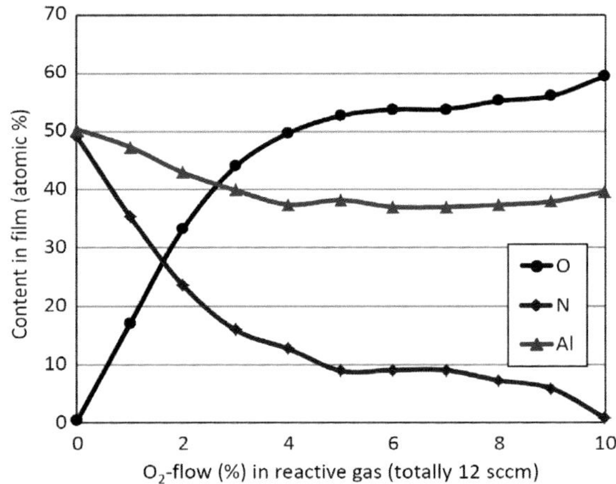

***Fig. 2:** Relative contents of N, O and Al in films produced with a gas mixture of N_2 and O_2.*

This information can be obtained in a single measurement of less than 10 minutes and is invaluable for this type of thin film development.

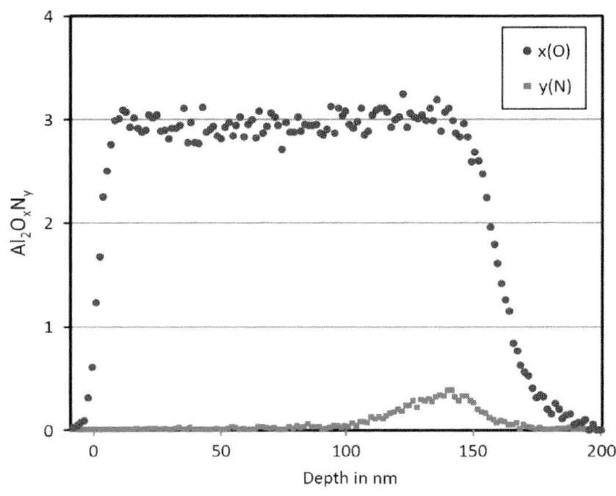

***Fig. 3:** Depth profile of O and N stoichiometric coefficients for an Al_2O_3 film grown at 10 % O_2.*

---

[1] *Nanoscale Materials Science, EMPA Dübendorf*

SUBSITUTION LEVEL OF THERMOELECTRIC OXIDES BY RBS

Materials for waste-heat recovery in solid oxide fuel cells

P. Thiel[1], S. Populoh[1], M. Döbeli

Thermoelectric converters can generate electrical energy when subjected to a thermal gradient. It allows the use of waste-heat to increase the efficiency of thermal processes. Within the framework of the HITTEC (High-temperature thermoelectric converters) project funded by CCEM and BFE our team at EMPA is developing such devices for the implementation into solid oxide fuel cells (SOFC) of the company Hexis (Winterthur).

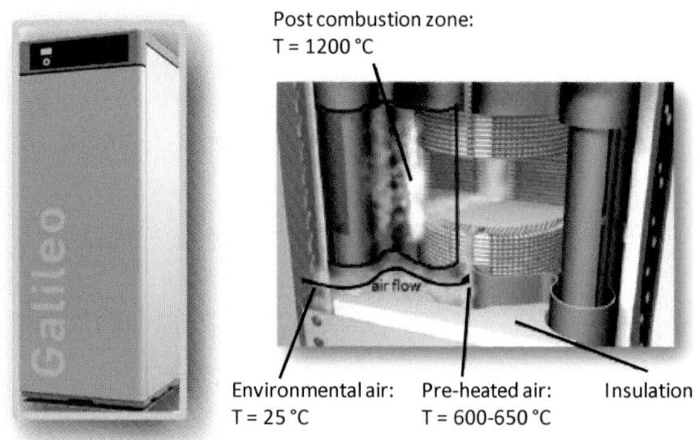

Fig. 1: *Solid oxide fuel cell system from Hexis.*

Oxide materials such as aliovalently substituted calcium manganese oxide ($CaMn_{1-x}A_xO_{3-\delta}$) are promising candidates as they can sustain the extreme conditions (>900°C in air) in a SOFC. To improve the performance the material is substituted with tungsten for adjusting the carrier concentration and reducing the thermal conductivity.

Fig. 2: *Thermoelectric oxide converters.*

The transport properties of $CaMn_{1-x}A_xO_{3-\delta}$ are primarily determined by the ratio of Mn^{3+}/Mn^{4+} ions and complete charge compensation can be assumed. With RBS the composition of the samples was determined and in combination with reduction experiments the Mn^{3+}/Mn^{4+} ratios calculated. Thanks to the accurate stoichiometry we were able to show that variations in the Seebeck coefficient are primarily due to the concentration of negative charge carriers. The high-temperature limit values from Mott's adiabatic small-polaron conduction model are in good agreement with Heikes formula for the Seebeck coefficient.

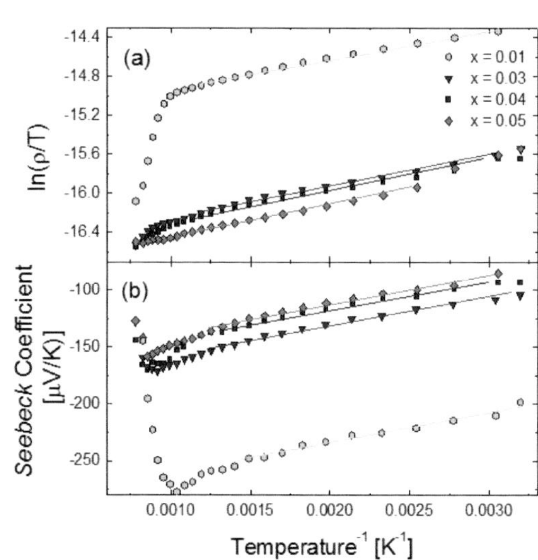

Fig. 3: *1/T plot of different $CaMn_{1-x}A_xO_{3-\delta}$ compounds.*

Considering as well the electrical conductivity a description of transport with the Cutler-Mott model is the most accurate.

[1] P. Thiel, S. Populoh et al., J. Appl. Phys. 114 (2013) 243707

[1] *Materials for Energy Conversion, EMPA Dübendorf*

EDUCATION

^{14}C dating of the 2014 growth season

Quaternary dating methods: Not only lectures

^{14}C DATING OF THE 2014 GROWTH SEASON

Measuring the atmospheric ^{14}C content from Zurich and Grenoble

I. Hajdas, A. Gogas[1], A. Weber[1], J. Braakhekke[2] S. Ivy-Ochs

During the one week visit at our laboratory high school students had an opportunity to gain some insights into the research and teaching activities at ETH. This year's project focused on the present day atmospheric ^{14}C content. In ^{14}C analyses reconstruction of the variable ^{14}C of the past atmosphere is an important part of radiocarbon dating (calibration) and it has been studied intensively in the past decades [1]. The present atmospheric ^{14}C content is also monitored and the development of the last decades is constantly updated [2]. However, the regional differences caused by different levels of pollution with CO_2 released due to fossil fuel combustion can be important when dating is applied to objects younger than 50 years, for example in forensic studies. Organic matter from two locations was sampled including a site in France visited by participants of an international workshop (Fig. 1).

Fig. 1: Location near Laffrey (France) where a hazel nut was found in the middle of an LGM till outcrop (see text). Sample location on the quarry wall shown by right hand of person in blue.

Leaves of trees were picked from the forest at Hönggerberg, Zurich with the intention to reconstruct the 2014 atmospheric ^{14}C composition. Additionally, a contemporary sample came from a surprising location. A hazel nut was found in a borrow on the wall of the glacial sediments located near a lake dammed by Last Glacial Maximum LGM frontal moraines of a tongue of the Romanche glacier (Fig. 2). The nut was presumably 'implanted' by birds and suspected to be of modern age though illustrating potential sources of 'reversals' in geochronology. All the dates were in a good agreement (Fig. 2).

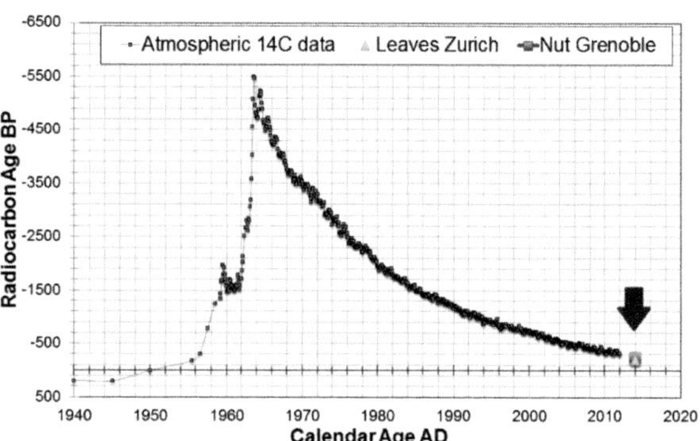

Fig. 2: Results of ^{14}C analyses of leaves and hazel nut all from 2014 showing a very good agreement and adding to atmospheric data [2].

In summary: this successful research project, which actively supported ^{14}C laboratory studies, illustrates the importance of learning by doing.

[1] I. Hajdas, *Treatise on Geochemistry (2nd Edition)* (2014) 37

[2] Q. Hua et al., Radiocarbon 55 (2013) 2059

---

[1] *Kantonsschule, Olten*

[2] *Geology, ETH Zurich*

QUATERNARY DATING METHODS: NOT ONLY LECTURES

Visit to an archaeological site in Otelfingen (Canton Zurich)

I. Hajdas, S. Ivy-Ochs

During the Quaternary Dating Methods lecture at the Earth Science Department of ETH Zurich, given in the fall-winter term, bachelor and master students have the chance to learn the basics of dating methods used in Quaternary research. These include: radiocarbon dating, in situ cosmogenic nuclide dating methods, U-Th, $^{40}Ar/^{39}Ar$ and other methods.

Fig. 1: *Rescue excavations at the construction site in Otelfingen. Evidence was found for several periods of occupation.*

Occasionally, we have the opportunity to take students to an active archeological site. For example, in 2009, we visited the Zurich Opera House site where abundant rests covering the last 5000 years were documented. This year we visited the archeological excavations in Otelfingen; a site that has been occupied since the Bronze Age. Several periods of occupation have been recognized, especially during Roman times and the medieval period. Otelfingen lies in Furttal along the southern foot of Lägern (860 m a.s.l.), the easternmost member of the folded Jura Mountains. Silex tool fragments found during excavation likely stem from Lägern. Along the crest of Lägern silex mining pits have been recognized.

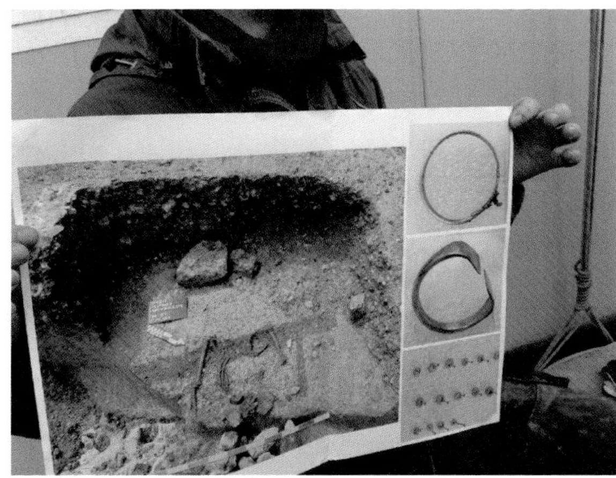

Fig. 2: *Archaeologists provide information for the public about the excavations at the site. Here pictures of spectacular finds made in summer 2014.*

Fig. 3: *Student taking a closer look at the sediments at the Otelfingen site in 2014.*

[1] K. Altorfer, Archäologie der Schweiz 33 (2010) 14

PUBLICATIONS

N. Akcar, S. Ivy-Ochs, V. Alfimov, A. Claude, H.R. Graf, A. Dehnert, P.W. Kubik, M. Rahn, J. Kuhlemann and C. Schlüchter
The first major incision of the Swiss Deckenschotter landscape
Swiss Journal of Geosciences **107** (2014) 337-347

N. Akcar, S. Ivy-Ochs, P. Deline, V. Alfimov, P.W. Kubik, M. Christl and C. Schlüchter
Minor inheritance inhibits the calibration of the Be-10 production rate from the AD 1717 Val Ferret rock avalanche, European Alps
Journal of Quaternary Science **29** (2014) 318-328

N. Akçar, V. Yavuz, S. Ivy-Ochs, R. Reber, P.W. Kubik, C. Zahno and C. Schlüchter
Glacier response to the change in atmospheric circulation in the eastern Mediterranean during the Last Glacial Maximum
Quaternary Geochronology **19** (2014) 27 - 41

A. Amorosi, L. Bruno, V. Rossi, P. Severi and I. Hajdas
Paleosol architecture of a late Quaternary basin–margin sequence and its implications for high-resolution, non-marine sequence stratigraphy
Global and Planetary Change **112** (2014) 12 - 25

T. Bekaddour, F. Schlunegger, H. Vogel, R. Delunel, K.P. Norton, N. Akcar and P.W. Kubik
Paleo erosion rates and climate shifts recorded by Quaternary cut-and-fill sequences in the Pisco valley, central Peru
Earth and Planetary Science Letters **390** (2014) 103-115

N. Bellin, V. Vanacker and P.W. Kubik
Denudation rates and tectonic geomorphology of the Spanish Betic Cordillera
Earth and Planetary Science Letters **390** (2014) 19-30

R. Belmaker, M. Stein, J. Beer, M. Christl, D. Fink and B. Lazar
Beryllium isotopes as tracers of Lake Lisan (last Glacial Dead Sea) hydrology and the Laschamp geomagnetic excursion
Earth and Planetary Science Letters **400** (2014) 233-242

A. Brauer, I. Hajdas, S.P.E. Blockley, C.B. Ramsey, M. Christl, S. Ivy-Ochs, G.E. Moseley, N.N. Nowaczyk, S.O. Rasmussen, H.M. Roberts, C. Spötl, R.A. Staff and A. Svensson
The importance of independent chronology in integrating records of past climate change for the 60e8 ka INTIMATE time interval
Quaternary Science Reviews (2014) 1 - 20

M.W. Buechi, F. Kober, S. Ivy-Ochs, B. Salcher, P.W. Kubik and M. Christl
Denudation rates of small transient catchments controlled by former glaciation: The Hörnli nunatak in the northeastern Swiss Alpine Foreland
Quaternary Geochronology **19** (2014) 135-147

U. Buentgen, L. Wacker, K. Nicolussi, M. Sigl, D. Guetler, W. Tegel, P.J. Krusic and J. Esper
Extraterrestrial confirmation of tree-ring dating
Nature Climate Change **4** (2014) 404-405

N. Casacuberta, M. Christl, J. Lachner, M.R. van der Loeff, P. Masque and H.-A. Synal
A first transect of U-236 in the North Atlantic Ocean
Geochimica Et Cosmochimica Acta **133** (2014) 34-46

J. Chen, M. Döbeli, D. Stender, K. Conder, A. Wokaun, C.W. Schneider and T. Lippert
Plasma interactions determine the composition in pulsed laser deposited thin films
Applied Physics Letters **105** (2014)

J.K. Chen, A. Palla-Papavlu, Y.L. Li, L.D. Chen, X. Shi, M. Döbeli, D. Stender, S. Populoh, W.J. Xie, A. Weidenkaff, C.W. Schneider, A. Wokaun and T. Lippert
Laser deposition and direct-writing of thermoelectric misfit cobaltite thin films
Applied Physics Letters **104** (2014)

H. Chittenden, R. Delunel, F. Schlunegger, N. Akcar and P.W. Kubik
The influence of bedrock orientation on the landscape evolution, surface morphology and denudation (Be-10) at the Niesen, Switzerland
Earth Surface Processes and Landforms **39** (2014) 1153-1166

M. Christl, X.X. Dai, J. Lachner, S. Kramer-Tremblay and H.-A. Synal
Low energy AMS of americium and curium
Nuclear Instruments & Methods in Physics Research Section B-Beam Interactions with Materials and Atoms **331** (2014) 225-232

M. Christl, R. Wieler and R.C. Finkel
Measuring one atom in a million billion with mass spectrometry
Elements **10** (2014) 330-332

A. Claude, S. Ivy-Ochs, F. Kober, M. Antognini, B. Salcher and P.W. Kubik
The Chironico landslide (Valle Leventina, southern Swiss Alps): age and evolution
Swiss Journal of Geosciences **107** (2014) 273-291

S. Das, K. Sen, I. Marozau, M.A. Uribe-Laverde, N. Biskup, M. Varela, Y. Khaydukov, O. Soltwedel, T. Keller, M. Döbeli, C.W. Schneider and C. Bernhard
Structural, magnetic, and superconducting properties of pulsed-laser-deposition-grown $La_{1}.85Sr_{0}.15CuO_{4}/La_{2}/3Ca_{1}/3MnO_{3}$ *superlattices on (001)-oriented* $LaSrAlO_{4}$ *substrates*
Physical Review B **89** (2014)

B. Dietre, C. Walser, K. Lambers, T. Reitmaier, I. Hajdas and J.N. Haas
Palaeoecological evidence for Mesolithic to Medieval climatic change and anthropogenic impact on the Alpine flora and vegetation of the Silvretta Massif (Switzerland/Austria)
Quaternary International **353** (2014) 3-16

D.Tikomiror, N. Akcar, S.Ivy-Ochs, V.Alfimov and C.Schlüchter
Calculation of shielding factors for production of cosmogenic nuclides in fault scarps
Quaternary Geochronology **19** (2014) 181-93

M. Döbeli, A. Dommann, X. Maeder, A. Neels, D. Passerone, H. Rudigier, D. Scopece, B. Widrig and J. Rammd
Surface layer evolution caused by the bombardment with ionized metal vapor
Nuclear Instruments and Methods in Physics Research B **332** (2014) 337 - 340

L. Fave, M.A. Pouchon, M. Döbeli, M. Schulte-Borchers and A. Kimura
Helium ion irradiation induced swelling and hardening in commercial and experimental ODS steels
Journal of Nuclear Materials **445** (2014) 235 - 240

L. Fave, M.A. Pouchon, M. Döbeli, M. Schulte-Borchers and A. Kimura
Experimental study of irradiation induced swelling in commercial and experimental ODS steels implanted with helium ions
European Nuclear Conference (2014) 212-219

C.J. Fogwill, C.S.M. Turney, N.R. Golledge, D.H. Rood, K. Hippe, L. Wacker, R. Wieler, E.B. Rainsley and R.S. Jones
Drivers of abrupt Holocene shifts in West Antarctic ice stream direction determined from combined ice sheet modelling and geologic signatures
Antarctic Science **26** (2014) 674-686

H. Galinski, T. Ryll, L. Yang, B. Scherrer, A. Evans, L.J. Gauckler and M. Döbeli
Platinum-based nanowire networks with enhanced oxygen-reduction activity
Physical Review Applied **2** (2014) 054015 (7 pp.)-054015 (7 pp.)

M. Gierga, M.P.W. Schneider, D.B. Wiedemeier, S.Q. Lang, R.H. Smittenberg, I. Hajdas, S.M. Bernasconi and M.W.I. Schmidt
Purification of fire derived markers for mu g scale isotope analysis (delta C-13, Delta C-14) using high performance liquid chromatography (HPLC)
Organic Geochemistry **70** (2014) 1-9

C. Glotzbach, M. Roettjer, A. Hampel, R. Hetzel and P.W. Kubik
Quantifying the impact of former glaciation on catchment-wide denudation rates derived from cosmogenic Be-10
Terra Nova **26** (2014) 186-194

S. Graca, V. Trabadelo, A. Neels, J. Kuebler, V. Le Nader, G. Gamez, M. Döbeli and K. Wasmer
Influence of mosaicity on the fracture behavior of sapphire
Acta Materialia **67** (2014) 67-80

C. Haeggi, R. Zech, C. McIntyre, M. Zech and T.I. Eglinton
On the stratigraphic integrity of leaf-wax biomarkers in loess paleosols
Biogeosciences **11** (2014) 2455-2463

N. Haghipour, J.-P. Burg, S. Ivy-Ochs, I. Hajdas, P.W. Kubik and M. Christl
Correlation of fluvial terraces and temporal steady-state incision on the onshore Makran accretionary wedge in southeastern Iran: Insight from channel profiles and ^{10}Be exposure dating of strath terraces
Geological Society of America Bulletin (2014) B31048. 1

I. Hajdas
Radiocarbon: Calibration to Absolute Time Scale
Treatise on Geochemistry (Second Edition), Elsevier, H.D.H.K. Turekian (2014) 37-43

I. Hajdas, C. Cristi, G. Bonani and M. Maurer
Textiles and radiocarbon dating
Radiocarbon **56** (2014) 637-643

O. Heiri, K.A. Koinig, C. Spötl, S. Barrett, A. Brauer, R. Drescher-Schneider, D. Gaar, S. Ivy-Ochs, H. Kerschner, M. Luetscher, A. Moran, K. Nicolussi, F. Preusser, R. Schmidt, P. Schoeneich, C. Schwörer, T. Sprafke, B. Terhorst and W. Tinner
Palaeoclimate records 60–8 ka in the Austrian and Swiss Alps and their forelands
Quaternary Science Reviews **106** (2014) 186-205

C. Herzog, J. Steffen, E.G. Pannatier, I. Hajdas and I. Brunner
Nine Years of Irrigation Cause Vegetation and Fine Root Shifts in a Water-Limited Pine Forest
Plos One **9** (2014)

K. Hippe, S. Ivy-Ochs, F. Kober, J. Zasadni, R. Wieler, L. Wacker, P.W. Kubik and C. Schlüchter
Chronology of Lateglacial ice flow reorganization and deglaciation in the Gotthard Pass area, Central Swiss Alps, based on cosmogenic Be-10 and in situ C-14
Quaternary Geochronology **19** (2014) 14-26

K. Hippe and N.A. Lifton
Calculating isotope ratios and nuclide concentrations for in situ cosmogenic C-14 analyses
Radiocarbon **56** (2014) 1167-1174

S. Ivy-Ochs, N. Akcar and A.J.T. Jull
Tracking the pace of Quaternary landscape change with cosmogenic nuclides
Quaternary Geochronology **19** (2014) 1-3

S. Ivy-Ochs and J.P. Briner
Dating Disappearing Ice With Cosmogenic Nuclides
Elements **10** (2014) 351-356

T. Jaeger, B. Bissig, M. Döbeli, A.N. Tiwari and Y.E. Romanyuk
Thin films of SnO2:F by reactive magnetron sputtering with rapid thermal post-annealing
Thin Solid Films **553** (2014) 21-25

J.P. Klages, G. Kuhn, C.D. Hillenbrand, A.G.C. Graham, J.A. Smith, R.D. Larter, K. Gohl and L. Wacker
Retreat of the West Antarctic Ice Sheet from the western Amundsen Sea shelf at a pre- or early LGM stage
Quaternary Science Reviews **91** (2014) 1-15

J. Lachner, M. Christl, V. Alfimov, I. Hajdas, P. W. Kubik, T. Schulze-König, L. Wacker and H.-A. Synal
^{41}Ca, ^{14}C and ^{10}Be concentrations in coral sand from the Bikini atoll
Journal of Environmental Radioactivity **129** (2014) 68 - 72

J. Lachner, M. Christl, A.M. Müller, M. Suter and H.-A. Synal
Be-10 and Al-26 low-energy AMS using He-stripping and background suppression via an absorber
Nuclear Instruments & Methods in Physics Research Section B-Beam Interactions with Materials and Atoms **331** (2014) 209-214

V. Maderich, K.T. Jung, R. Bezhenar, G. de With, F. Qiao, N. Casacuberta, P. Masque and Y.H. Kim
Dispersion and fate of Sr-90 in the Northwestern Pacific and adjacent seas: Global fallout and the Fukushima Dai-ichi accident
Science of the Total Environment **494** (2014) 261-271

X. Maeder, M. Döbeli, A. Dommann, A. Neels, H. Rudigier, B. Widrig and J. Ramm
Phase formation in cathodic arc synthesized Al–Hf and Al–Hf–O coatings during high temperature annealing in ambient air
Surface and Coatings Technology **260** (2014) 56-62

I. Marozau, P.T. Das, M. Döbeli, J.G. Storey, M.A. Uribe-Laverde, S. Das, C.N. Wang, M. Rossle and C. Bernhard
Influence of La and Mn vacancies on the electronic and magnetic properties of $LaMnO_3$ thin films grown by pulsed laser deposition
Physical Review B **89** (2014)

S. Martin, P. Campedel, S. Ivy-Ochs, A. Vigano, V. Alfimov, C. Vockenhuber, E. Andreotti, G. Carugati, D. Pasqual and M. Rigo
Lavini di Marco (Trentino, Italy): Cl-36 exposure dating of a polyphase rock avalanche
Quaternary Geochronology **19** (2014) 106-116

C. Münsterer, L. Wacker, B. Hattendorf, M. Christl, J. Koch, R. Dietiker, H.-A. Synal and D. Guentherk
Rapid Revelation of Radiocarbon Records with Laser Ablation Accelerator Mass Spectrometry
Chimia **68** (2014) 215-216

T. Prokscha, H. Luetkens, E. Morenzoni, G.J. Nieuwenhuys, A. Suter, M. Doebeli, M. Horisberger and E. Pomjakushina
Depth dependence of the ionization energy of shallow hydrogen states in ZnO and CdS
Physical Review B **90** (2014)

A. Quiles, H. Valladas, J. Geneste, J. Clottes, D. Baffier, B. Berthier, F. Brock, C.B. Ramsey, E. Delqué-Količ, J. Dumoulin, I. Hajdas, K. Hippe, G.W.L. Hodgins, A. Hogg, A.J.T. Jull, E. Kaltnecker, M.d. Martino, C. Oberlin, F. Petchey, P. Steier, H.-A. Synal, J.v.d. Plicht, E.M. Wild and A. Zazzo
Second Radiocarbon Intercomparison Program for the Chauvet-Pont d'Arc Cave, Ardèche, France
Radiocarbon **56** (2014) 1-18

S.O. Rasmussen, H.H. Birks, S.P.E. Blockley, A. Brauer, I. Hajdas, W.Z. Hoek, J.J. Lowe, A. Moreno, H. Renssen, D.M. Roche, A.M. Svensson, P. Valdes and M.J.C. Walker
Dating, synthesis, and interpretation of palaeoclimatic records of the Last Glacial cycle and model-data integration: advances by the INTIMATE (INTegration of Ice-core, MArine and TErrestrial records) COST Action ES0907
Quaternary Science Reviews **106** (2014) 1-13

R. Reber, N. Akcar, S. Ivy-Ochs, D. Tikhomirov, R. Burkhalter, C. Zahno, A. Luethold, P.W. Kubik, C. Vockenhuber and C. Schlüchter
Timing of retreat of the Reuss Glacier (Switzerland) at the end of the Last Glacial Maximum
Swiss Journal of Geosciences **107** (2014) 293-307

R. Reber, N. Akcar, S. Yesilyurt, V. Yavuz, D. Tikhomirov, P.W. Kubik and C. Schlüchter
Glacier advances in northeastern Turkey before and during the global Last Glacial Maximum
Quaternary Science Reviews **101** (2014) 177-192

S. Savi, K. Norton, V. Picotti, F. Brardinoni, N. Akcar, P.W. Kubik, R. Delunel and F. Schlunegger
Effects of sediment mixing on Be-10 concentrations in the Zielbach catchment, central-eastern Italian Alps
Quaternary Geochronology **19** (2014) 148-162

S. Savi, K.P. Norton, V. Picotti, N. Akcar, R. Delunel, F. Brardinoni, P.W. Kubik and F. Schlunegger
Quantifying sediment supply at the end of the last glaciation: Dynamic reconstruction of an alpine debris-flow fan
Geological Society of America Bulletin **126** (2014) 773-790

I. Schimmelpfennig, J.M. Schaefer, N. Akçar, T. Koffman, S. Ivy-Ochs, R. Schwartz, R.C. Finkel, S. Zimmerman and C. Schlüchter
A chronology of Holocene and Little Ice Age glacier culminations of the Steingletscher, Central Alps, Switzerland, based on high-sensitivity beryllium-10 moraine dating
Earth and Planetary Science Letters **393** (2014) 220-230

M.J. Simon, C.L. Zhou, M. Döbeli, A. Cassimi, I. Monnet, A. Méry, C. Grygiel, S. Guillous, T. Madi, A. Benyagoub, H. Lebius, A.M. Müller, H. Shiromaru and H.-A. Synal
Measurements and 3D Monte Carlo simulation of MeV ion transmission through conical glass capillaries
Nuclear Instruments and Methods in Physics Research B **330** (2014) 11-17

M.J. Simon, C.L. Zhou, M. Döbeli, T. Ikeda, A.M. Müller, A. Benyagoub, C. Grygiel, S. Guillous, H. Lebius, A. Mery, I. Monnet, F. Ropars, H. Shiromaru and A. Cassimi
Simulation of MeV ion transmission through glass micro-capillaries
Xxviii International Conference on Photonic, Electronic and Atomic Collisions (Icpeac) **488** (2014)

E. Stilp, A. Suter, T. Prokscha, Z. Salman, E. Morenzoni, H. Keller, C. Katzer, F. Schmidl and M. Döbeli
Modifications of the Meissner screening profile in $YBa_2Cu_3O_7\text{–}\delta$ thin films by gold nanoparticles
Physical Review **B 89** (2014) 1-5

S. Szidat, G.A. Salazar, E. Vogel, M. Battaglia, L. Wacker, H.-A. Synal and A. Türler
^{14}C Analysis and Sample Preparation at the New Bern Laboratory for the Analysis of Radiocarbon with AMS (LARA)
Radiocarbon **56** (2014) 561-566

V. Vanacker, N. Bellin, A. Molina and P.W. Kubik
Erosion regulation as a function of human disturbances to vegetation cover: a conceptual model
Landscape Ecology **29** (2014) 293-309

L. Wacker, D. Guettler, J. Goll, J.P. Hurni, H.-A. Synal and N. Walti
Radiocarbon dating to a single year by means of rapid atmospheric C-14 changes
Radiocarbon **56** (2014) 573-579

Y.-L. Zhang, J. Li, G. Zhang, P. Zotter, R.-J. Huang, J.-H. Tang, L. Wacker, A.S.H. Prevot and S. Szidat
Radiocarbon-based source apportionment of carbonaceous aerosols at a regional background site on Hainan Island, South China
Environmental Science & Technology **48** (2014) 2651-2659

B. Zollinger, C. Alewell, C. Kneisel, K. Meusburger, D. Brandová, P.W. Kubik, M. Schaller, M. Ketterer and M. Egli
The effect of permafrost on time-split soil erosion using radionuclides (^{13}7Cs, ^{239}+ ^{240}Pu, meteoric ^{10}Be) and stable isotopes ($\delta\ ^{13}$C) in the eastern Swiss Alps
Journal of Soils and Sediments (2014) 1-20

P. Zotter, I. El-Haddad, Y. Zhang, P.L. Hayes, X. Zhang, Y.-H. Lin, L. Wacker, J. Schnelle-Kreis, G. Abbaszade, R. Zimmermann, J.D. Surratt, R. Weber, J.L. Jimenez, S. Szidat, U. Baltensperger and A.S.H. Prevot
Diurnal cycle of fossil and nonfossil carbon using radiocarbon analyses during CalNex
Journal of Geophysical Research-Atmospheres **119** (2014) 6818-6835

TALKS AND POSTERS

N. Akçar, V. Yavuz, S. Ivy-Ochs
Investigation of the Quaternary geological context of the February 2011 massive failures at the Çöllolar coalfield, eastern Turkey
Turkey, Muğla, 13.10.2014, 8th International Symposium on Eastern Mediterranean Geology

N. Akçar, S. Ivy-Ochs, M. Christl, A. Claude, C. Wirsig, J. Lachner, S. Padilla
Optimization of cosmogenic ^{10}Be and ^{26}Al extraction for precise AMS measurements of low concentrations
France, Aix-en-Provence, 26.08.2014, AMS-13

N. Akçar, V. Yavuz, S. Ivy-Ochs, S. Yeşilyurt, R. Reber, D. Tikhomirov, P.W. Kubik, C. Vockenhuber, C. Schlüchter
Extensive glaciations in Anatolian Mountains during the global Last Glacial Maximum
Austria, Vienna, 27.04.2014, EGU General Assembly

V. Alfimov, S. Ivy-Ochs, P.W. Kubik, J. Beer, H.-A. Synal
^{36}Cl in 100-m long limestone core from Vue des Alpes, Switzerland:
When was the last major glaciation there?
France, Aix-en-Provence, 26.08.2014, AMS-13

N. Casacuberta, M. Christl, C. Vockenhuber, C. Walther, M. R. Van-der-Loeff, P. Masqué, H.-A. Synal
Distribution of ^{236}U, ^{129}I and ^{240}Pu/^{239}Pu ratios in Arctic Ocean waters
USA, Honolulu, 23.02.2014, Ocean Science Meeting 2014

N. Casacuberta, M. Christl, C. Vockenhuber, C. Walther, M. R. Van-der-Loeff, P. Masqué, H.-A. Synal
^{236}U and ^{129}I as tracers of water masses in the Arctic Ocean
Germany, Berlin, 20.03.2014, DPG Spring Meeting

N. Casacuberta, M. Christl, J. Lachner, M. R. Van-der-Loeff, P. Masqué, C. Vockenhuber, C. Walther, H.-A. Synal
^{236}U, ^{129}I and Pu-isotopes as oceanographic tracers in the Arctic and Atlantic Ocean
France, Aix-en-Provence, 26.08.2014, AMS-13

N. Casacuberta, M. Christl, C. Vockenhuber, M. Gorny, C. Walther, H.-A. Synal
Radiochemistry of ^{236}U, ^{129}I and Pu-isotopes in seawater samples
France, Aix-en-Provence, 26.08.2014, AMS-13

M. Christl, B. Chayeron, N. Casacuberta, H.-A. Synal
^{236}U and Pu-isotopes in two corals from French Polynesia
France, Aix-en-Provence, 26.08.2014, AMS-13

M. Christl, J. Lachner, N. Casacuberta, C. Vockenhuber, V. Alfimov, H.-A. Synal, I. Goroncy, J. Herrmann
The distribution of ^{236}U and ^{129}I in the North Sea in 2009
France, Aix-en-Provence, 26.08.2014, AMS-13

M. Christl, X. Dai, S. Kramer-Tremblay, N. Priest, J. Lachner, H.-A. Synal, X. Zhao, L. Kieser, A.E. Litherland
Improved Target Preparation Methods for Actinides by AMS
France, Aix-en-Provence, 26.08.2014, AMS-13

M. Christl, S. Maxeiner, J. Lachner, C. Vockenhuber, A.M. Müller, N. Casacuberta, H.-A. Synal
Towards a compact multi isotope AMS system – status and applications
Germany, Berlin, 20.03.2014, DPG Spring Meeting

M. Christl
Compact, low energy accelerator mass spectrometry (AMS) – a new tool for the ultra-sensitive detection of actinides in environmental samples
Switzerland, Lausanne, 21.11.2014, IRA Lausanne

A. Claude, N. Akçar, S. Ivy-Ochs, F. Schlunegger, P.W. Kubik, A. Dehnert, M. Rahn, C. Schlüchter
Timing of Early Quaternary Glaciations in the Alps
Austria, Vienna, 27.04.2014, EGU General Assembly

A. Claude, N. Akçar, S. Ivy-Ochs, F. Schlunegger, P.W. Kubik, A. Dehnert, M. Rahn, C. Schlüchter
Timing of Early and Middle Quaternary Glaciations in the Alps
Switzerland, Fribourg, 22.11.2014, 12th Swiss Geoscience Meeting

A. Daraoui, M. Schwinger, B. Riebe, C. Walther, C. Vockenhuber, H.-A. Synal
Application of AMS for determination of I-129 in soil profiles from the northern and southern hemisphere
Germany, Berlin, 20.03.2014, DPG Spring Meeting

B. Riebe, C. Walther, C. Vockenhuber, H.-A. Synal
Iodine isotopes (^{127}I and ^{129}I) at the Zugspitze
Germany, Hannover, 18.11.2014, IRS

B. Dittmann, T. Dunai, A. Dewald, S. Heinze, C. Feuerstein, E. Strub, K. Fifeld, M. Srncik, S. Tims, A. Wallner, H.-A. Synal, M. Christl
Development of a new reference material for isotopic ratio measurements of plutonium with AMS
France, Aix-en-Provence, 26.08.2014, AMS-13

M.J. Simon, M. Döbeli, A.M. Müller, M. Schulte-Borchers, H.-A. Synal
Simulation of MeV ion transmission through capillaries
USA, San Antonio, 28.05.2014, CAARI-23

M. Döbeli
Ion-solid interaction
Switzerland, Zurich, 24.06.2014, AK FIB Workshop

M. Döbeli
Formation of micro-beams by capillary collimation
Italy, Legnaro, 03.07.2014, Joint Training Course on Ion Beam Microscopy

M. Döbeli
Ionizing radiation – safety and protection
Switzerland, Zurich, 08.10.2014, LIP AMS Seminar

M. Döbeli
Ion-Solid Interaction for Secondary Ion Mass Spectrometry (SIMS)
Switzerland, Zurich, 05.11.2014, CCMX Advanced Course

R. Dobrowolski, J. Rodzik, P. Mroczek, P. Zagórski, K. Bałaga, I.Hajdas, M. Wołoszyn, M. Piotrowski
Hoards from a wetland (Cherven Towns, Eastern Poland), Czermno – an archaeological and paleogeographical approach
Switzerland, Bern, 11.06.2014, Culture, Climate and Environment Interactions at Prehistoric Wetland Sites.

S. Fahrni
Radiocarbon measurements at the University of California, Irvine
Switzerland, Zurich, 30.04.2014, LIP AMS Seminar

S. Fahrni, J. Park, B. Fuller, M. Friedrich,R. Muscheler, L. Wacker, J. Southon, R. E. Taylor
Annual german oak and bristlecone pine ^{14}C data sets confirm 2625 BP ^{14}C wiggle and reveal fine structure
France, Aix-en-Provence, 26.08.2014, AMS-13

S. Fahrni , L. Wacker, C. McIntyre, M. Seiler, P. Gautschi, J. Bourquin
Advances and applications in ^{14}C gas analysis at ETH Zurich
Switzerland, Zurich, 17.09.2014, LIP AMS Seminar

S. Fahrni, L. Wacker, M. Moros, L. Zillen-Snowballd, T. Tuna, Y. Fagault, E. Bard
Developments and applications in 14C gas analysis: dating a sediment core with sub-mg foraminifera samples
France, Aix-en-Provence, 25.08.2014, AMS-13

C. R. Frew R. Pellitero, M. Spagnolo, B. R. Rea, J. Bakke, P. D. Hughes, S. Ivy-Ochs, S. Lukas, H. Renssen, A. Ribolini
Automating the implementation of an equilibrium profile model for glacier reconstruction in a GIS environment
Austria, Vienna, 27.04.2014, EGU General Assembly

I. Hajdas, B. Giaccio, R. Isaia, S. Nomade
New Radiocarbon Ages for Campanian Ignimbrite (Italy) eruption at 40 ka
Spain, Zaragoza, 15.06.2014, INTIMATE Open Workshop and COST Action ES0907 Final Event

I. Hajdas, A. Fontana, L. Hendriks, K. Hippe, S. Ivy-Ochs
Evaluation of ^{14}C preparation methods for chronologies of LGM records
Austria, Innsbruck, 27.09.2014, DEUQUA 2014

I. Hajdas
Radiocarbon analysis and Quaternary Geochronology
Austria, Vienna, 27.04.2014, EGU General Assembly

I. Hajdas
AMS Radiocarbon dating at ETH Zurich, prospects for mortar
France, Bordeaux, 04.06.2014, Séminaire Au Pied du Mur : Architecture de l'Antiquité tardive au Moyen Âge
I. Hajdas, G. Bonani, S. Fahrni, M. Maurer, C. McIntyre, M. Belen Roldan Torres de Roettig
Radiocarbon dating of mortar at ETH Zurich
Italy, Padua, 14.04.2014, The Third Mortar Dating Workshop

I. Hajdas
C-14 methodology : recent developments in the ETH laboratory - Zürich
France, Dieppe, 04.09.2014, Reunion FALEME

I. Hajdas
Recent developments in ^{14}C analyses and applications
Switzerland, Basel, 26.11.2014, Current topics in Geosciences , University of Basel

K. Hippe, A. Fontana, I. Hajdas, S. Ivy-Ochs
High-resolution radiocarbon chronology of the LGM stratigraphy of the Cormor alluvial megafan (Tagliamento glacier, NE Italy)
Spain, Zaragoza, 15.06.2014, INTIMATE Open Workshop and COST Action ES0907 Final Event

K. Hippe, I. Hajdas, S. Ivy-Ochs, M. Maisch
Chronologie der Klimaschwankungen des Mittleren Würm (MIS3) im nördlichen Alpenvorland
Austria, Innsbruck, 27.09.2014, DEUQUA 2014

K. Hippe, I. Hajdas, S. Ivy-Ochs, M. Maisch
Chronology of Middle Würm climate changes in the Swiss Alpine foreland
Switzerland, Fribourg, 22.11.2014, 12th Swiss Geoscience Meeting

K. Hippe
In situ cosmogenic 14C in Quaternary Geochronology and Earth surface process studies
UK, Edinburgh, 28.11.2014, Hutton Club Seminar

S. Ivy-Ochs, C. Wirsig, M. Christl, C. Vockenhuber
Using cosmogenic ^{10}Be and ^{36}Cl to assess the depth of glacial erosion in the Alps
Denmark, Aarhus, 13.06.2014, Nordic Workshop on Cosmogenic Nuclide Dating

S. Ivy-Ochs
Overview of the timing of deglaciation on the northern Alpine forelands
France, Grenoble-Chambéry, 12.11.2014, INQUA CECLAP Workshop LGM in the Alps 2014

S. Ivy-Ochs, S. Martin
Dating large landslides in northern Italy
Italy, Padua, 08.04.2014, University of Padua Earth Science seminar series

S. Ivy-Ochs
Evaluating the timing of Younger Dryas glacier advances in Europe
Italy, Sambuco, 30.09.2014, Leverhulme network meeting

D. Jenson, C. Vockenhuber, M. Adamic, J. Olson, M. Watrous
Iodine Standard Materials : Preparation and Inter-Laboratory Comparisons
France, Aix-en-Provence, 26.08.2014, AMS-13

R.S. Jones, A.N. Mackintosh, K.P. Norton, N.R. Golledge, C.J. Fogwill, P.W. Kubik, M. Christl and S.L. Greenwood
Rapid thinning of an East Antarctic outlet glacier during stable Holocene climate
USA, San Francisco, 19.12.2014, AGU Fall Meeting

R.S. Jones, A.N. Mackintosh, K.P. Norton, N.R. Golledge, C.J. Fogwill, P.W. Kubik
Glacial history and behaviour of Mackay Glacier, Transantarctic Mountains
New Zealand, Auckland, 27.08.2014, SCAR

J.Lachner, J. Beer, M. Christl, M. Stockhecke
Das Laschamp-Event mittels ^{10}Be im Vansee
Germany, Berlin, 20.03.2014, DPG Spring Meeting

M. Lupker, K. Hippe, F. Kober, L. Wacker, R. Wieler
Sediment residence time in a Himalayan catchment: insights from combined in-situ ^{14}C and ^{10}Be measurements in river sands
France, Aix-en-Provence, 23.08.2014, Pre-AMS-13 workshop

S. Mandal, M. Lupker, N. Haghipour, J.-P. Burg, M. Christl
Landscape development in Southern Peninsular India from 10Be denudation rates in river sands
Austria, Vienna, 01.05.2014, EGU General Assembly

S. Maxeiner, M. Seiler, M. Suter, H.-A. Synal
Charge state distributions and charge exchange cross sections of carbon in helium at 45-260 keV
France, Aix-en-Provence, 26.08.2014, AMS-13

S. Maxeiner, M. Suter, M. Christl, H.-A. Synal
Simulation of ion beam scattering in a gas stripper
France, Aix-en-Provence, 26.08.2014, AMS-13

S. Maxeiner, M. Suter, M. Christl, H.-A. Synal
Simulationen zur Bestimmung von Gasverteilung und Projektilstreuung im Gasstripper
Germany, Berlin, 20.03.2014, DPG Spring Meeting

S. Maxeiner
Optimizing low energy and compact Accelerator Mass Spectrometry systems
Switzerland, Zurich, 12.09.2014, Zurich PhD seminar 2014

S. Maxeiner
Gasdichten in differentiell gepumpten Wechselwirkungszonen
Switzerland, Brunegg, 12.06.2014, GV Schweizerische Vakuumgesellschaft (SVG)

S. Maxeiner, A. Herrmann, A.M. Müller, M. Suter, C. Vockenhuber, H.-A. Synal
Developments towards a 300 kV multi-isotope AMS system
Switzerland, Zurich, 24.09.2014, LIP AMS Seminar

C. McIntyre, L. Wacker, S. Fahrni, T. Eglinton
Simultaneous and precise ^{13}C and ^{14}C measurements of gas samples
France, Aix-en-Provence, 26.08.2014, AMS-13

C. McIntyre, F. Lechleitner, S. Lang, L. Wacker, S. Fahrni, T. Eglinton
^{14}C Contamination Testing Using Wet Chemical Oxidation and a Gas Ion Source
France, Aix-en-Provence, 26.08.2014, AMS-13

C. McIntyre
Simultaneous ^{13}C and ^{14}C analysis
UK, Manchester, 14.10.2014, Isoprime Ltd

A. Moran ,S. Ivy-Ochs, H. Kerschner
Alpine Glacier Oscillations and Climate in the Early Holocene
Austria, Vienna, 27.04.2014, EGU General Assembly

A. Moran, S. Ivy-Ochs, H. Kerschner
Re-evaluation of the moraines in the Kromer Valley (Silvretta Mountains, Austria)
Austria, Innsbruck, 26.09.2014, DEUQUA 2014

N. Mozafari Amiri, Ö Sümer, D. Tikhomirov, Ç Özkaymak, S. Ivy-Ochs, B. Uzel, C. Vockenhuber, H. Sözbilir, N. Akçar
Holocene seismic activity and slip rates of the Priene-Sazlı Fault, Western Anatolia
Turkey, Muğla, 13.10.2014, 8th International Symposium on Eastern Mediterranean Geology

N. Mozafari Amiri, Ö Sümer, D. Tikhomirov, Ç Özkaymak, S. Ivy-Ochs, B. Uzel, C. Vockenhuber, H. Sözbilir, N. Akçar
Reconstructing the paleoseismic history of the Priene-Sazli Fault using 36Cl cosmogenic nuclide dating method, Western Anatolia, Turkey
Austria, Vienna, 27.04.2014, EGU General Assembly

A. M. Müller, M. Döbeli, H.-A. Synal
Recent developments in AMS at ETH Zurich
France, Obernai, 02.10.2014, IBAF 2014

A. M. Müller, M. Döbeli, H.-A. Synal
Recent results with the CAEN DPP
France, Obernai, 01.10.2014, IBAF 2014

A. M. Müller
Development of high resolution gas ionization detectors
Switzerland, Zurich, 09.04.2014, LIP AMS Seminar

A. M. Müller, M. Döbeli, M. Seiler, H.-A. Synal
A new simplified Bragg type detector for AMS and IBA applications
France, Aix-en-Provence, 26.08.2014, AMS-13

A. M. Müller, M. Christl, J. Lachner, S. Maxeiner, C. Zanella, H.-A. Synal
Al-26 measurements below 500 kV
France, Aix-en-Provence, 26.08.2014, AMS-13

A. Ojeda, C. W. Schneider, M. Döbeli, T. Lippert, A. Wokaun
Angular distribution of species in pulsed laser deposition of La(x)Ca(1-x)MnO(3)
France, Lille, 28.05.2014, EMRS Spring Meeting

R. Pellitero, C. Frew, M. Spagnolo, B. R. Rea, J. Bakke, P. D. Hughes, S. Ivy-Ochs, S. Lukas, H. Renssen, A. Ribolini
Automating the implementation of an equilibrium profile model for glacier reconstruction and Equilibrium Line Altitude calculation in a GIS environment
UK, Glasgow, 16.04.2014, 22nd GIS Research UK

B. Scherrer, H. Galinski, M. Döbeli, J. Cairney
Oxygen transport in noble metals
Germany, Stuttgart, 02.09.2014, Atom Probe Tomography & Microscopy 2014

C. Schlüchter, N. Akçar, V. Yavuz, M. Leuenberger, S. Ivy-Ochs, R. Reber, D. Tikhomirov, C. Yeşilyurt, C. Zahno, P.W. Kubik
Quaternary glaciations in Anatolia - potential correlations and implications
Turkey, Muğla, 13.10.2014, 8th International Symposium on Eastern Mediterranean Geology

J. Schoonejans, V. Vanacker, S. Opfergelt, Y. Ameijeiras-Mariño, P.W. Kubik
Coupling of physical erosion and chemical weathering after phases of intense human activity.
Austria, Vienna, 29.04.2014, EGU General Assembly

M.Schulte-Borchers, M. Döbeli, M. Klöckner, A.M. Müller, A. Eggenberger, M.J. Simon, H.-A. Synal
Use of a capillary microprobe for heavy ion microbeams in Ion Beam Analysis and MeV SIMS
Italy, Padova, 06.07.2014, ICNMTA Microbeam Conference

M. Seiler, S. Maxeiner, H.-A. Synal
Messstabilität an der myCADAS Anlage
Germany, Berlin, 20.03.2014, DPG Spring Meeting

M. Seiler, S. Maxeiner, L. Wacker, H.-A. Synal
Status on mass spectrometric radiocarbon detection at ETHZ
France, Aix-en-Provence, 27.08.2014, AMS-13

M. Seiler, S. Fahrni, P. Gautschi, C. McIntyre, L. Wacker, H.-A. Synal
Direct injection of carbon dioxide from headspace vials into a gas ion source
France, Aix-en-Provence, 28.08.2014, AMS-13

M. Seiler
Technical improvements at the myCADAS
Switzerland, Zurich, 21.05.2014, LIP AMS Seminar

U. Sojc, I. Hajdas, S. Ivy-Ochs, N. Akçar, P. Deline
Building high resolution radiocarbon dating chronologies for the reconstruction of late Holocene landslide events in the Mont Blanc area, Italy
Switzerland, Fribourg, 22.11.2014, 12th Swiss Geoscience Meeting

E. Strub, H.H. Coenen, U. Herpers, H. Wiesel, G. Delisle, S. Binnie, J.T. Dunai, A. Liermann, A Dewald, C. Feuerstein, M. Christl
Glaciation history of Queen Maud Land (Antarctica) - new exposure data from Nunataks
France, Aix-en-Provence, 26.08.2014, AMS-13

M. Suter
From nuclear physics instruments to modern AMS
Austria, Vienna, 28.11.2014, Symposium zum 75. Geburtstag von Prof. Walter Kutschera

M. Suter
Different views of accelerator mass spectrometry
Switzerland, Dübendorf, 16.10.2014, Symposium to Honor Prof Dr. Jürg Beer

M. Suter
Siliziumfolien für die Trennung von Isobaren - einige grundsätzliche Überlegungen
Germany, Berlin, 20.03.2014, DPG Spring Meeting

M. Suter
What can we learn from modeling the physics of AMS?
France, Aix-en-Provence, 27.07.2014, AMS-13

H.-A. Synal, M. Christl, S. Maxeiner, A.M. Müller, J. Lachner, M. Seiler, M. Suter, Ch. Vockenhuber, L. Wacker
Progress in Design and Performance of Low Energy AMS Systems: Status Report
France, Aix-en-Provence, 26.08.2014, AMS-13

H.-A. Synal, S. Maxeiner, M. Seiler
MASS SPECTROMETRIC C-14 DETECTION TECHNIQUES: Progress Report
UK, Ware, 12.06.2014, Glaxo-Smith-Kline

H.-A. Synal
Analytical challenges inspired by the demand of visionary applications
Switzerland, Dübendorf, 16.10.2014, Workshop in Honor of Jürg Beer

H.-A. Synal
Latest Developments in Accelerator Mass Spectrometry
South Korea, Anyang, 04.12.2014, Workshop in Bioplastic Materials

H.-A. Synal
Progress in Radiocarbon detection: Applications and Selected Examples
South Korea, Anyang, 05.12.2014, Korea Apparel Testing & Research Institute

H.-A. Synal
The Art of Dating: 14C-Altersbestimmungen, neue Nachweismethoden und ausgewählte Beispiele
Switzerland, Dübendorf, 15.01.2014, Rotary-Club Meeting

H.-A. Synal
Wie alt ist Mona Lisa? Atome lügen nicht...
Switzerland, Zurich, 30.11.2014, ETH Zürich Science City publich lecures

C. Terrizzano, R. Zech, E. García Morabito, M. Yamin, N. Haghipour, L. Wüthrich, M. Christl, S. Willett, J. M. Cortes, A. E. Rapalini, V.A. Ramos
Timing of neotectonic deformation in the western Precordillera, Central Andes of Argentina, inferred from ^{10}*Be Surface Exposure Dating*
Austria, Innsbruck, 26.09.2014, DEUQUA 2014

V. Vanacker, N. Bellin, J. Schoonejans, A. Molina, P.W. Kubik
Sensitivity of mountain ecosystems to human-accelerated soil erosion. Contrasting geomorphic response between tropical and semi-arid ecosystems.
Austria, Vienna, 01.05.2014, EGU General Assembly

V. Vanacker, J. Schoonejans, N. Bellin, Y. Ameijeiras-Mariño, S. Opfergelt, M. Christl
Detecting anthropogenic disturbance on weathering and erosion processes
USA, San Francisco, 17.12.2014, AGU Fall Meeting

V. Vanacker, J. Schoonejans, N. Bellin, P.W. Kubik
Human disturbances to soil systems in the Western Mediterranean
UK, Edinburgh, 17.01.2014, Invited seminar, Hutton Club

V. Vanacker, J. Schoonejans, N. Bellin, P.W. Kubik
Human disturbances to soil systems in the Western Mediterranean
USA, Tempe, 09.12.2014, Arizona State University, SESE Group

C. Vockenhuber, M. Thšni, J. Jense), K. Arstila, J. Julin, H. Kettunen, M.I. Laitinen, M. Rossi, T. Sajavaara, H.J. Whitlow, O. Osmani, A. Schinner, P. Sigmund
Structure in the velocity dependence of heavy-ion energy-loss straggling
Hungary, Debrecen, 13.07.2014, ICACS Hungary

C. Vockenhuber
AMS of ^{129}*I: cross contamination and its correction*
Germany, Berlin, 20.03.2014, DPG Spring Meeting

C. Vockenhuber, T. Schulze-Kšnig, H.-A. Synal, I. Aeberli, M. B. Zimmermann
Efficient ^{41}*Ca measurements for biomedical applications*
France, Aix-en-Provence, 26.08.2014, AMS-13

C. Vockenhuber, N. Casacuberta, M. Christl, H.-A. Synal
^{129}*I towards its lower limits*
France, Aix-en-Provence, 26.08.2014, AMS-13

C. Vockenhuber, K.-U. Miltenberger, M. Suter, H.-A. Synal
Isobar-separation techniques for 6 MV Tandem accelerators
France, Aix-en-Provence, 26.08.2014, AMS-13

L. Wacker, J. Bourquin, S. Fahrni, C. McIntyre, H.-A. Synal
Gas Ion Source + Automated Graphitization Equipment
UK, Ware, 12.06.2014, Glaxo-Smith-Kline

L. Wacker, J. Beer, S. Bollhalder, U. Büntgen, M. Friedrich, B. Kromer, D. Nievergelt, H.-A. Synal
Improving the tree-based radiocarbon calibration curve
France, Aix-en-Provence, 26.08.2014, AMS-13

L. Wacker, S. Fahrni, H.-A. Synal
Challenges in Radiocarbon analysis: Dating smaller samples more accurate
Romania, Bukarest, 03.12.2014, Seminar in Magurele

L. Wacker, J. Beer, U. Büntgen, D. Güttler, B. Kromer, H.-A. Synal
Fast changes in atmospheric radiocarbon concentrations in the past
18.08.2014, Radiocarbon in the environment

A. Wallner, T. Belgya, M. Bichler, K. Buczak, M. Christl, I. Dillmann Iris, K. Fifeld, M. Hotchkis, F. Kappele, A. Krasa, J. Lachner, J. Lippol A. Plompen, F. Quinto, V. Semkova, M. Srncik, P. Stei L. Szentmiklosi, S. Tims, S. Winkler
A novel method for studying neutron-induced reactions on actinides
France, Aix-en-Provence, 26.08.2014, AMS-13

M. Watrous, M. Adamic, J. Delmore, R. Hague, D. Jenson, J. Olson, C. Vockenhuber
Installation of a 0.5 MV Accelerator Mass Spectrometry at Idaho National Laboratory
France, Aix-en-Provence, 26.08.2014, AMS-13

M. Watrous, M. Adamic, T. Lister, J. Olson, D. Jenson, C. Vockenhuber
Electrodeposition as an alternate method for preparation of environmental samples for iodide for analysis by AMS
France, Aix-en-Provence, 26.08.2014, AMS-13

C. Münsterer, L. Wacker, B. Hattendorf, M. Christl, J. Koch, R. Dietiker, D. Günther, H.-A. Synal,
Radiocarbon Analyses by Laser Ablation
Switzerland, Zurich, 26.02.2014, LIP AMS Seminar

C. Münsterer, L. Wacker, B. Hattendorf, J. Koch, M. Christl, R. Dietiker, D. Günther, H.-A. Synal
Laserablation-AMS: zeitsparende online-^{14}C-Analyse von Karbonaten
Germany, Berlin, 20.03.2014, DPG Spring Meeting

C. Münsterer, L. Wacker, B. Hattendorf, M. Christl, J. Koch, R. Dietiker, D. Günther, H.-A. Synal,
Rapid ^{14}C-Analysis by Laser Ablation- Accelerator Mass Spectrometry
Germany, Heidelberg, 24.04.2014, Instituts Seminar IUP

C. Münsterer, L. Wacker, B. Hattendorf, M. Christl, J. Koch, R. Dietiker, H.-A. Synal, D. Günther
Rapid Radiocarbon Analyses by Laser Ablation
France, Aix-en-Provence, 27.08.2014, AMS-13

C. Wirsig, S. Ivy-Ochs
When did the ice disappear?
Switzerland, Parpan, 24.05.2014, Phd Retreat D-ERDW

C. Wirsig, S. Ivy-Ochs, M. Christl, C. Vockenhuber, J. Reitner, M. Bichler, M. Reindl, C. Schlüchter, H.-A. Synal
Quantifizierung glazialer Erosion mit kosmogenen Nukliden
Germany, Berlin, 20.03.2014, DPG Spring Meeting

C. Wirsig, S. Ivy-Ochs, M. Christl, J. Reitner, M. Reindl, M. Bichler, C. Vockenhuber, N. Akcar, C. Schlüchter
Constraining local subglacial bedrock erosion rates with cosmogenic nuclides
Austria, Vienna, 01.05.2014, EGU General Assembly

C. Wirsig, S. Ivy-Ochs, J. Zasadni, M. Christl, N. Akcar, J. Reitner, C. Vockenhuber, C. Schlüchter
Refining the deglaciation chronology in the High Alps based on cosmogenic Be-10
Denmark, Aarhus, 12.06.2014, Nordic Workshop on Cosmogenic Nuclide Dating

C. Wirsig, S. Ivy-Ochs, M. Christl, J. Reitner, M. Reindl, M. Bichler, C. Vockenhuber, P.W. Kubik, H.-A. Synal, C. Schlüchter
Combined cosmogenic Be-10 and Cl-36 nuclide concentrations constrain subglacial erosion rates
France, Aix-en-Provence, 27.08.2014, AMS-13

C. Wirsig
Mit Butter hobelt man nicht!
Switzerland, Zurich, 17.12.2014, LIP AMS Seminar

H. Wittmann, N. Dannhaus, F. Blanckenburg, J. Bouchez, A. Suessenberger, J. Guyot, L. Maurice, N. Filizola, J. Gaillardet and M. Christl
Reactive and dissolved meteoric ^{10}Be/^{9}Be ratios in the Amazon basin
Austria, Vienna, 01.05.2014, EGU General Assembly

L. Wüthrich, R. Zech, N. Haghipour, C. Gnägi, M. Christl, S. Ivy-Ochs, H. Veit
Extremely eroded or incredibly young - ^{10}Be depth profile dating of moraines in the Swiss Midlands
Austria, Vienna, 02.05.2014, EGU General Assembly

C. Zabcı, T. Sançar, D. Tikhomirov, C. Vockenhuber, S. Ivy-Ochs, N. Akçar
Understanding intra-plate deformation in Anatolia: Insights from preliminary slip-rates of the Malatya-Ovacık Fault, Eastern Turkey, during the last 16 ka
Turkey, Muğla, 13.10.2014, 8th International Symposium on Eastern Mediterranean Geology

C. Zabcı, T. Sançar, D. Tikhomirov, C. Vockenhuber, S. Ivy-Ochs, N. Akçar
Preliminary geologic slip rates of the Ovacik Segment (Malatya-Ovacik Fault, Turkey) for the last 15 ka: Insights from cosmogenic ^{36}Cl dating of offset fluvial surfaces
Austria, Vienna, 01.05.2014, EGU General Assembly

G. Zeilinger, F. Kober, K. Hippe, O. Marc
Tectonically controlled denudation rates in the central Bolivian Andes
Austria, Vienna, 02.05.2014, EGU General Assembly

R. Zurfluh, F. Kober, S. Ivy-Ochs, I. Hajdas, M. Christl
Post-glacial landscape evolution of the upper Haslital Aare between Handegg and Guttannen (Bernese Alps)
Switzerland, Fribourg, 22.11.2014, 12th Swiss Geoscience Meeting

SEMINAR

Spring semester

19.02.2014
Ulla Heikkilä (EAWAG, Switzerland), How do atmospheric and climatic influences hamper the use of ice core ^{10}Be as solar activity proxy?

26.02.2014
Caroline Münsterer (ETHZ, Switzerland), Radiocarbon analyses by Laser Ablation

05.03.2014
Lorenz Wüthrich (ETHZ, Switzerland), Extremely eroded or incredibly young – ^{10}Be depth profile dating of moraines in the Swiss Midlands

12.03.2014
Michael Wiedenbeck (GFZ Potsdam, Germany), Large Geometry SIMS Instruments in the Geosciences

19.03.2014
Philipp Steinmann (BAG, Switzerland), Stratosphären-Tracer (Be-7 und Na-22) in bodennaher Luft: Ergebnisse aus der Langzeitüberwachung der Radioaktivität in der Luft

26.03.2014
Pascal Froidevaux (CHUV, Switzerland), Use of radiometric methods in forensic science at CHUV: the Arafat's case

02.04.2014
Mandal Sanjay Kumar (ETHZ, Switzerland), Low denudation recorded by ^{10}Be in river sands from the Southern Peninsular India

09.04.2014
Timothy Eglinton (ETHZ, Switzerland), New biogeochemical insights from radiocarbon-rich data sets

16.04.2014
Arnold Müller (ETHZ, Switzerland), Development of gas ionization detectors

23.04.2014
Shasta Marrero (University of Edinburgh, Great Britain), Ice Sheet History in the Ellsworth Mountains of Antarctica Using Cosmogenic Nuclides

30.04.2014
Simon Fahrni (ETHZ, Switzerland), Radiocarbon measurements at University of California Irvine

07.05.2014
Martin Martschini (VERA, Austria), First photodetachment experiments at the Ion-Laser-Interaction-Setup (ILIAS) in Vienna and their relevance for AMS

14.05.2014
Olivia Kronig (ETHZ, Switzerland), Late Holocene Evolution of the Triftjeglacier Constrained with ^{10}Be Exposure and Radiocarbon Dating

14.05.2014
Rafael Zurfluh (ETHZ, Switzerland), Post-glacial Evolution of the Upper Haslital Aare between Handegg and Guttannen

21.05.2014
Martin Seiler (ETHZ, Switzerland), Technical improvements at the ultra compact radiocarbon system myCADAS

28.05.2014
Tobias Jocham and Michael Marx (Berlin-Brandenburgische Akademie der Wissenschaften, Germany), Zeitliche und räumliche Einordnung früher Qur'ānhandschriften – Möglichkeiten und Grenzen der Paläographie

04.06.2014
Klaus-Ulrich Miltenberger (ETHZ, Switzerland), Evaluation of Mg-26 suppression by post-accelerator foil stripping for Al-26 measurements using AlO- ions at the Zürich 6 MV Tandem AMS facility

04.06.2014
Claudia Zanella (ETHZ, Switzerland), Al-26 Measurements with Low Energy Accelerator Mass Spectrometry

11.06.2014
Georg Steinhauser (Colstate University, USA), Monitoring of Environmental Radionuclides after the Fukushima Accident and Comparison with Chernobyl

Fall semester

17.09.2014
Phillip Gautschi (ETHZ, Switzerland) and Simon Fahrni (ETHZ, Switzerland), Semesterarbeit: New direct gas interface for the MICADAS system

24.09.2014
Sascha Maxeiner (ETHZ, Switzerland), Towards a 300 kV multi-isotope AMS system

01.10.2014
Francesca Quinto (KIT, Germany), AMS of Actinides in Groundwater: Development of a New Procedure for Simultaneous Trace Analysis of U, Np, Pu, Am and Cm Isotopes

08.10.2014
Max Döbeli (ETHZ, Switzerland), Ionizing radiation – safety and protection

15.10.2014
Amade Bortis (ETHZ, Switzerland), Semesterarbeit: Experiments on prefocussing a capilary ion beam

22.10.2014
Kristina Hippe (ETHZ, Switzerland), Middle Würm radiocarbon chronologies (50-25 ka) in the Alpine foreland: results from the TiMIS project

29.10.2014
Markus Küffner (SIK-ISEA, Switzerland) and Laura Hendriks (ETHZ, Switzerland), On analytical studies of paintings and importance of microsize ^{14}C-dating

05.11.2014
Ola Fredin (Quaternary Geology, Norway), Ice dynamics and deglaciation of SW Norway constrained by LIDAR mapping and cosmogenic nuclide exposure ages of glacial landforms

12.11.2014
Matthias George (ETHZ, Switzerland), Digital pulse processing

12.11.2014
Amade Bortis (ETHZ, Switzerland), Semesterarbeit: Experiments on prefocussing a capilary ion beam

19.11.2014
Bao Rui (ETHZ, Switzerland), Hydrodynamic control on organic carbon dispersal and burial in shallow marginal seas

26.11.2014
Franziska Lechleitner (ETHZ, Switzerland), Stalagmite ^{14}C as proxy for solar activity and climate? Insights from a tropical cave

03.12.2014
Maxi Castrillejo (UAB, Spain), Reassessment of ^{90}Sr, ^{137}Cs and ^{134}Cs in the Pacific Ocean and the coast off Japan derived from the Fukushima Dai-ichi nuclear accident

10.12.2014
Christoph Rembser (CERN, Switzerland), It's a gas, gas, gas! - (Gas filled particle detectors - current applications and trends for the future)

17.12.2014
Christian Wirsig (ETHZ, Switzerland), Mit Butter hobelt man nicht!

THESES (INTERNAL)

Term papers

Philip Gautschi
A new direct gas interface system for radiocarbon analysis
ETH Zurich

Klaus-Ulrich Miltenberger
Evaluation of ^{26}Mg suppression by post-accelerator foil stripping for ^{26}Al measurements using AlO- ions at the Zurich 6 MV EN Tandem AMS facility
ETH Zurich

Claudia Zanella
Aluminum Measurements with low energy mass spectrometry
ETH Zurich

Doctoral theses

Martin Seiler
Accelerator Mass Spectrometry for Radiocarbon at very low Energies
ETH Zurich

THESES (EXTERNAL)

Term papers

Laura Hendriks
Radiocarbon dating of the stratigraphic record of Belinga (NE Italy) and associated ca. 35 ka old tree trunks
ETH Zurich (Switzerland)

Laura Hinrichsen
Preparation of parchment, papyrus and wood for "Corpus Coranicum"
Berlin-Brandenburgischen Akademie, Berlin (Germany)

Monika Isler
Preparation of wood for archeological samples
University of Zurich (Switzerland)

Alban Kolly
Analysis of the Quaternary deposits in the Senn gravel pit near Edlisbach Kt. Zug
ETH Zurich (Switzerland)

Simon Parolo
Beryllium Enrichment at the Surface by performing Heat Treatments with a Cu-Be Alloy and Depth Profiling using the ERDA Technique
ETH Zurich (Switzerland)

Diploma/Master theses

Henrik Becker
Experimental Study of the tritium production by neutron activation in the solid deuterium moderator of the ultra-cold neutron source at PSI
ETH Zurich (Switzerland)

Jan Casagrande
Erosion rates by ^{10}Be measurement of a postglacial setting in the Central Alps of Switzerland, Flühli/Sörenberg in Emmental
University of Bern (Switzerland)

Veit Dausmann
Radiogenic Isotope Evolution of the Deep Arctic Ocean since the Late Miocene: Ferromanganese Crust Evidence
GEOMAR Helmholtz-Zentrum für Ozeanforschung Kiel (Germany)

Laura Hendriks
Microscale radiocarbon dating of paintings
ETH Zurich (Switzerland)

Olivia Kronig
Late Holocene evolution of the Triftjegletscher constrained with ^{10}Be exposure and radiocarbon dating
ETH Zurich (Switzerland)

Sieber Matthias
Investigating the radiocarbon characteristics of covalently-bound components of sedimentary organic matter
ETH Zurich (Switzerland)

Kündig Nicole
Composition and provenance of terrestrial biomarkers within the Inn River drainage basin
ETH Zurich (Switzerland)

Fabio Nunes
Where are the major sediment sources of Val Lumnezia (GR, eastern Swiss Alps)?: Quantitative estimate of cosmogenic ^{10}Be-deived erosion rates and sediment discharges
University of Bern (Switzerland)

Wenk Pascal
Grain size-specific organic carbon isotope stratigraphy of Iberian margin sediments
ETH Zurich (Switzerland)

Ursula Sojc
Building high-resolution radiocarbon chronologies for the reconstruction of the late Holocene glacier variations and landslide events in the Mont Blanc area, Italy
ETH Zurich (Switzerland)

Raphael Zurfluh
Post-glacial landscape evolution of the upper Haslital Aare between Handegg and Guttannen (Bernese Alps, Switzerland)
ETH Zurich (Switzerland)

Doctoral theses

Nicolas Bellin
Human impact on soil erosion based on ^{10}Be cosmogenic radionuclide measurements in a semi-arid region : the case of the Betic Ranges
Université catholique de Louvain, Louvain-la-Neuve (Belgium)

Voss Britta
Spatial and temporal dynamics of biogeochemical processes in the Fraser River, Canada: A coupled organic-inorganic perspective
Massachusetts Institute of Technology and Woods Hole Oceanographic Institution (USA)

Jikun Chen
Analysis of Laser-Induced Plasmas Utilizing $^{18}O_2$ as Oxygen Tracer
Paul Scherrer Institut and ETH Zurich (Switzerland)

Müller Claudia
Phase separation in co-sputtered immiscible Cu- Ta alloy films
ETH Zurich (Switzerland)

Bernoulli Daniel
Cohesive & Adhesive Failure and Contact Damage of Diamond-Like Carbon (DLC) Coated Titanium Substrates
ETH Zurich (Switzerland)

Marie Guns
Sediment dynamics in tropical mountain regions: influence of anthropogenic disturbances on sediment transfer mechanisms
Université catholique de Louvain, Louvain-la-Neuve (Belgium)

Daniel Muff
Development of Mechanically Robust Esthetic Coating Systems for Dental Implants
ETH Zurich (Switzerland)

Christina Pecnik
Esthetics and Reliability of Thin Films for Dental Implants
ETH Zurich (Switzerland)

Amandine Perret
Géopatrimoines des troic Chablais: identification et valorisation des témoins glaciaires
University of Lausanne (Switzerland)

Regina Reber
Late Quaternary glaciations in eastern Anatolia: Moraine records in Kaçkar Mountain Valleys (first mapping and cosmogenic nuclide dating) with a methodlogical example from central Switzerland
University of Bern (Switzerland)

Philipp Reibisch
Low-Dimensional Compounds and Composites for Lithium Exchange as well as for Electronic and for Ionic Conductivity Enhancements
ETH Zurich (Switzerland)

Annika Mareike Schwinger
Deutschlandweite Untersuchung der ^{129}I- und ^{127}I-Inventare und ihres Isotopenverhältnisses in verschiedenen Umweltkompartimenten
Uni Hannover (Germany)

Shuqin Tao
The composition, isotopic characterization and sources of organic matter in the Yellow River suspended particulates and adjacent Bohai and Yellow Sea surface sediments
ETH Zurich (Switzerland) and Ocean University of China (China)

Dmitry Tikhomirov
An advanced model for fault scarp dating and paleoearthquake reconstruction, with a case study of the Gediz Graben formation (Turkey)
University of Bern (Switzerland)

COLLABORATIONS

Australia

University of Sidney, Australian Centre for Microscopy & Microanalysis, Sidney

The Australian National University, Department of Nuclear Physics, Canberra

Austria

AlpS - Zentrum für Naturgefahren- und Riskomanagement GmbH, Geology and Mass Movements, Innsbruck

Geological Survey of Austria, Sediment Geology, Vienna

University of Innsbruck, Institutes of Geography, of Geology, and of Botany, Innsbruck

University of Vienna, VERA, Faculty of Physics, Vienna

Vienna University of Technology, Institute for Geology, Vienna

Belgium

Royal Institute for Cultural Heritage, Brussels

Université catholique de Louvain, Earth and Life Institute– TECLIM, Louvain-la-Neuve

Canada

Chalk River Laboratories, Dosimetry Services, Chalk River

TRIUMF, Vancouver

University of Ottawa, Department of Earth Sciences, Ottawa

China

Chinese Academy of Sciences, Institute of Botany, Beijing

China Earthquake Administration, Beijing

Peking University, Accelerator Mass Spectrometry Lab., Beijing

Denmark

Danfysik, A/S, Taastrup

Risø DTU, Risø National Laboratory for Sustainable Energy, Roskilde

Technical University of Denmark, Department of Photonics Engineering , Roskilde

University of Southern Denmark, Department of Physics, Chemistry and Pharmacy, Odense

Finland

University of Jyväskylä, Physics Department, Jyväskylä

France

Aix-Marseille University, Collège de France, Aix-en-Provence

Commissariat à l'énergie atomique et aux énergies alternatives, Laboratoire des Sciences du Climat et de l'Environnement (LSCE), Gif-sur-Yvette Cedex

Laboratoire de biogeochimie moléculaire, Strasbourg

Université de Savoie, Laboratoire EDYTEM, Le Bourget du Lac

Germany

Alfred Wegener Institute of Polar and Marine Research, Marine Geochemistry, Bremerhaven

BSH Hamburg, Radionuclide Section, Hamburg

Deutsches Bergbau Museum, Bochum

GFZ German Research Centre for Geosciences, Earth Surface Geochemistry and Dendrochronology Laboratory, Potsdam

Helmholtz-Zentrum Dresden-Rossendorf, DREAMS, Rossendorf

Hydroisotop GmbH, Schweitenkirchen

IFM-GEOMAR, Departments of Palaeo-Oceanography and Biologcal-Oceanography, Kiel

Leibniz-Institut für Ostseeforschung Warnemünde, Marine Geologie, Rostock

Regierungspräsidium Stuttgart, Landesamt für Denkmalpflege, Esslingen

Marum, Micropalaeontology - Paleoceanography, Bremen

Reiss-Engelhorn-Museum, Klaus-Tschira-Labor, Mannheim

Senckenberg am Meer, Deutsches Zentrum für Marine Biodiversitätsforschung, Wilhelmshaven

University of Cologne, Deoartments of Geology, of Mineralogy, and of Physics, Cologne

University of Hannover, Institute for Radiation Protection and Radioecology, Hannover

University of Heidelberg, Institute of Environmental Physics, Heidelberg

University of Hohenheim, Institute of Botany, Stuttgart

University of Münster, Institute of Geology and Paleontology, Münster

University of Tübingen, Department of Geosciences, Tübingen

Hungary

Hungarian Academy of Science, Institute of Nuclear Research (ATOMKI), Debrecen

India

Inter-University Accelerator Center, Accelerator Division, New Dehli

Israel

Hebrew University, Geophysical Institute of Israel, Jerusalem

Italy

CAEN S.p.A., Viareggio

CNR Rome, Institute of Geology , Rome

Geological Survey of the Provincia Autonoma di Trento, Trento

INGV Istituto Nazionale di Geofisica e Vulcanologia, Sez. Sismologia e Tettonofisica, Rome

University of Bologna, Deptartment Earth Sciences, Bologna

University of Padua, Department of Geology and Geophysics, Padua

University of Turin, Department of Geology, Turin

Japan

University of Tokai, Department of Marine Biology, Tokai

Liechtenstein

OC Oerlikon AG, Balzers

Mexico

UNAM (Universidad Nacional Autonoma de Mexico), Instituto de Fisica, Mexico

New Zealand

University of Waikato, Radiocarbon Dating Laboratory, Waikato

Victoria University of Wellington, School of Geography, Environment and Earth Sciences, Wellington

Norway

Norwegian University of Science and Technology, Physical Geography, Trondheim

University of Bergen, Department of Earth Science and Department of Biology, Bergen

University of Norway, The Bjerkness Centre for Climate Researcu, Bergen

Poland

University of Marie Curie Sklodowska, Department of Geography , Lublin

Romania

Horia Hulubei - National Institute for Physics and Nuclear Engineering, , Magurele

Singapore

National University of Singapore, Department of Chemistry, Singapore

Slovakia

Comenius University, Faculty of Mathematics, Physics and Infomatics, Bratislava

Slovenia

Geological Survey of Slovenia, Ljubljana

South Korea

Korea Apparel Testing & Research Institute, Seoul

Spain

Autonomous University of Barcelona, Environmental Science and Technology Institute, Barcelona

University of Murcia, Department of Plant Biology, Murcia

University of Seville, National Center for Accelerators, Seville

Sweden

University of Uppsala, Angström Institute, Uppsala

Lund University, Department of Earth and Ecosystem Sciences, Lund

Switzerland

ABB Ltd, Lenzburg

Centre Hospitalier Universitaire Vaudois, Institut de Radiophysique, Lausanne

Dendrolabor Wallis, Brig

Empa, Research groups: Nanoscale Materials Science, Solid State Chemistry, Advanced Materials Processing, Functional Polymers, Joining Technology and Corrosion, and Thin Films, Dübendorf

Empa, Mechanics of Materials and Nanostructures, Thun

ENSI, Brugg

EPFL, Photovoltaics and Thin Films, Lausanne

ETH Zurich: Departments of Health Sciences, Materials, and Institutes of Geochemistry and Petrology, of Particle Physics, of Engineering Geology, of Geochemistry and Petrology, Isotope Geochemistry and Mineral Resources, and of Geology, Zurich

Evatec AG, Flums

Federal Office for Civil Protection, Spiez Laboratory, Spiez

Fine Arts Eexperts Institute, Geneva

Glas Trösch AG, Bützberg

Gübelin Gem Lab Ltd. (GGL), Luzern

Haute Ecole ARC, IONLAB, La-Chaux-de-Fonds

Helmut Fischer AG, Hünenberg

Kanton Bern, Achäologischer Dienst, Bern

Kanton Graubünden, Archäologischer Dienst, Chur

Kanton Solothurn, Kantonsarchäologie, Solothurn

Kanton Turgau, Kantonsarchäologie, Frauenfeld

Kanton Zug, Kantonsarchäologie, Zug

Kanton Zürich, Kantonsarchäologie, Dübendorf

Labor für quartäre Hölzer, Affoltern a. Albis

Laboratiore Romand de Dendrochronologie, Moudon

Landesmuseum, Zurich

Laserenterprise, Zürich

Office et Musée d'Archéologie Neuchatel, Neuchatel

Paul Scherrer Institut (PSI): Laboratories for Atmospheric Chemistry, for Radiochemistry and Environmental Chemistry, Materials Group, Muon Spin Rotation, Materials Science Beamline, Villigen

Research Station Agroscope Reckenholz-Tänikon ART, Air Pollution / Climate Group, Zurich

Stadt Zürich, Amt für Städtebau, Zurich

Swiss Federal Institute of Aquatic Science and Technology (Eawag), SURF, Dübendorf

Swiss Federal Institute for Forest, Snow and Landscape Reseach (WSL), Landscape Dynamics, Dendroecology, and Soil Sciences, Birmensdorf

Swiss Gemmological Institute - SSEF, Basel

Swiss Institute for Art Research, SIK ISEA, Zurich

University of Basel, Departement Altertumswissenschaften, Institut für Prähistorische und Naturwissenschaftliche Archäologie (IPNA), Basel

University of Bern, Oeschger Center for Climate Research, Institute of Geology, Berne

University of Geneva, Departments of Geology/Paleontology and of Anthropology/Ecology, Geneva

University of Neuchatel, Department of Geology, Neuchatel

University of Freiburg, Department of Physics, Faculty of Environmentat and Natural Resources, Freiburg

University of Zurich, Departments of Biochemistry, and of Geography, Institute of Geography, Abteilung Ur- und Frühgeschichte, Zürich

Turkey

Istanbul Technical University, Faculty of Mines, Istanbul

Tübitak, Marmara Arastirma Merkezi, Gebze Kocaeli

United Kingdom

Brithish Arctic Survey, Cambridge

Durham University, Department of Geography, Durham

Newcastle University, School for History, Classics and Archaeology , Newcastle

Northumbria University, Department of Geography, Newcastle

University of Bristol, Schools of Chemistry and of Earth Sciences, Bristol

University of Cambridge, Department of Earth Sciences, Cambridge

University of Oxford, Department of Earth Sciences, Oxford

USA

Colorado State University, Department of Environmental and Radiological Health Sciences, Fort Collins

Columbia University, LDEO, New York

Eckert & Ziegler Vitalea Science, AMS Laboratory, Davis

Florida State University, Earth, Ocean & Atmospheric Science, Tallahassee

Idaho National Laboratory, National and Homeland Security, Idaho Falls

Lamont-Doherty Earth Observatory, Department of Geochemistry, Palisades

University of Utah, Geology and Geophysics, Salt Lake City

Woods Hole Oceanographic Institution, Center for Marine and Environmental Radioactivity, Woods Hole

VISITORS AT THE LABORATORY

Jessica Castleton
University of Utah, Utah, USA
01.01.2014 - 31. 01.2014

Dinakar Kanjilal
Inter-University Accelerator Centre, New Delhi, India
20.01.2014 - 21. 01.2014

Devinder Mehta
Department of Physics, Panjab University, Chandigarh, India
20.01.2014 - 21. 01.2014

Ashok Kumar
Department of Physics, Panjab University, Chandigarh, India
20.01.2014 - 21. 01.2014

Silvana Martin
Geoscience Department, University of Padua, Padua, Italy
23.01.2014 - 27. 01.2014

Heinz Gäggeler
University of Bern, Bern, Switzerland
27.01.2014 - 27. 01.2014

John Olson
Idaho National Laboratory, Idaho Falls, USA
27.01.2014 - 31. 01.2014

Douglas D. Jenson
Idaho National Laboratory, Idaho Falls, USA
31.01.2014 - 31. 01.2014

Thibaut Tuna
Centre Européen de Recherche et d'Enseignement des Géosciences de l'Environnement, Aix-en-Provence, France
03.02.2014 - 06. 02.2014

Yoann Fagault
Centre Européen de Recherche et d'Enseignement des Géosciences de l'Environnement, Aix-en-Provence, France
03.02.2014 - 06. 02.2014

Edouard Bard
Centre Européen de Recherche et d'Enseignement des Géosciences de l'Environnement, Aix-en-Provence, France
03.02.2014 - 06. 02.2014

Christian Eisenach
Institute of Geology and Paleontology, University of Münster, Münster, Germany
05.02.2014 - 05. 02.2014

Adam Sookdeo
University of Ottawa, Ottawa, Canada
18.02.2014 - 18. 02.2014

Jens Fohlmeister
Environmental Physics, University of Heidelberg, Heidelberg, Germany
25.02.2014 - 25. 02.2014

Giulia Guidobaldi
Dipartimento di Scienzhe della Terra, University of Pisa, Pisa, Italy
01.03.2014 - 31. 05.2014

Michael Wiedenbeck
Helmholtz Zentrum Potsdam, Potsdam, Germany
12.03.2014 - 13. 03.2014

Tobias J. Jocham
Berlin-Brandenburgische Akademie der Wissenschaften, Potsdam, Germany
21.03.2014 - 21. 03.2014

Pascal Froidevaux
Université de Lausanne, Lausanne, Switzerland
26.03.2014 - 26. 03.2014

Stephanie Schneider
Institute for Radioecology and Radiation Protection, Leibniz University Hannover, Hannover, Germany
14.04.2014 - 16. 04.2014

Gerd Helle
GFZ German Reserach Centre for GeoSciences, Helmholtz Centre Potsdam, Potsdam, Germany
21.04.2014 - 21. 04.2014

Ulf Büntgen
Landscape Dynamics, Dendroökologie, Swiss Federal Research Institute WSL, Birmensdorf, Switzerland
21.04.2014 - 21. 04.2014

Shasta Marrero
University of Edinburgh, Edinburgh, Scotland
23.04.2014 - 25. 04.2014

Martin Martschini
Vienna Environmental Research Accelarator, University of Vienna, Vienna, Austria
07.05.2014 - 07. 05.2014

Karl Håkansson
Universität Uppsala, Uppsala, Sweden
19.05.2014 - 23. 05.2014

Jonas Åström
Universität Uppsala, Uppsala, Sweden
19.05.2014 - 23. 05.2014

Kook-Hyun Yu
Korea Institute of Radiological & Medical Sciences, Seoul, South Korea
19.05.2014 - 19. 05.2014

Jong Guk Kim
Korea Institute of Radiological & Medical Sciences, Seoul, South Korea
19.05.2014 - 19. 05.2014

Byoung Soo Kim
Korea Institute of Radiological & Medical Sciences, Seoul, South Korea
19.05.2014 - 19. 05.2014

Tobias J. Jocham
Berlin-Brandenburgische Akademie der Wissenschaften, Potsdam, Germany
28.05.2014 - 28. 05.2014

Julia Heuss
ETH Studienwoche, Physik ohne Grenzen, Zurich, Switzerland
02.06.2014 - 06. 06.2014

Riccarda Albertin
ETH Studienwoche, Physik ohne Grenzen, Zurich, Switzerland
02.06.2014 - 06. 06.2014

Fadrina Denoth
ETH Studienwoche, Physik ohne Grenzen, Zurich, Switzerland
02.06.2014 - 06. 06.2014

Noa Reber
ETH Studienwoche, Physik ohne Grenzen, Zurich, Switzerland
02.06.2014 - 06. 06.2014

Thomas Ronge
Geosciences, Marine Geology and Paleontology, Alfred-Wegener-Institut, Bremerhaven, Germany
02.06.2014 - 06. 06.2014

Georg Steinhauser
Colorado State University, Fort Collins, USA
11.06.2014 - 11. 06.2014

Alessandro Fontana
Geoscience Department, University of Padua, Padua, Italy
08.07.2014 - 10. 07.2014

Heinz Gäggeler
University of Bern, Bern, Switzerland
15.07.2014 - 15. 07.2014

Keith Fifield
Department of Nuclear Physics, Australian National University, Canberra, Australia
15.07.2014 - 15. 07.2014

Park Yong-Bin
Korea Evaluation Institute of Industrial Technology, Seoul, South Korea
22.08.2014 - 22. 08.2014

Leem Seungyoon
Korea Apparel Testing & Research Institute, Seoul, South Korea
01.09.2014 - 01. 09.2014

Lee Sanghyuk
Korea Apparel Testing & Research Institute, Seoul, South Korea
01.09.2014 - 01. 09.2014

Lee Eunae
Korea Apparel Testing & Research Institute, Seoul, South Korea
01.09.2014 - 01. 09.2014

Lena Hellmann
Landscape Dynamics, Dendroökologie, Swiss Federal Research Institute WSL, Birmensdorf, Switzerland
15.09.2014 - 18. 09.2014

Maxi Castrillejo
Universitat Autònom de Barcelona, Barcelona, Spain
22.09.2014 - 19. 12.2014

Johannes Lachner
Vienna Environmental Research Accelarator, University of Vienna, Vienna, Austria
23.09.2014 - 23. 09.2014

Francesca Quinto
Karlsruhe Institute of Technology, Karlsruhe, Germany
01.10.2014 - 01. 10.2014

Guillaume Soulet
Geology & Geophysics, Woods Hole Oceanographic Institution, Woods Hole, USA
01.10.2014 - 01. 10.2014

Ronny Friedrich
Klaus-Tschira-Archäometrie-Zentrum, Curt-Engelhorn-Zenrum Archäometrie gGmbH, Mannheim, Germany
22.10.2014 - 22. 10.2014

Bernd Kromer
Klaus-Tschira-Archäometrie-Zentrum, Curt-Engelhorn-Zenrum Archäometrie gGmbH, Mannheim, Germany
22.10.2014 - 22. 10.2014

Harald Müller
Institute for Materials Science and Authenticity Testing, Wiesbaden, Germany
30.10.2014 - 30. 10.2014

Ola Fredin
Norges geologiske undersøkelse (NGU), Trondheim, Norway
03.11.2014 - 05. 11.2014

Michèle Köhli
University of Zurich, Zurich, Switzerland
14.11.2014 - 14. 11.2014

Alexander Gogas
Kantonsschule Olten, Olten, Switzerland
17.11.2014 - 21. 11.2014

Andreas Weber
Kantonsschule Olten, Olten, Switzerland
17.11.2014 - 21. 11.2014

Tobias J. Jocham
Berlin-Brandenburgische Akademie der Wissenschaften, Potsdam, Germany
01.12.2014 - 01. 12.2014

Keith Fifield
Department of Nuclear Physics, Australian National University, Canberra, Australia
01.12.2014 - 02. 12.2014

Rusland Cusnir
Institut de Radiophysique, Lausanne, Switzerland
10.12.2014 - 11. 12.2014

Marc Ostermann
Geology Department, University of Innsbruck, Innsbruck, Austria
15.12.2014 - 19. 12.2014

Jordi Garcia-Orellana
Universitat Autònom de Barcelona, Barcelona, Spain
16.12.2014 - 17. 12.2014